Nadia ERRAFIY
Adnane MOUTAOUAKKIL

Oxidative and nitrosative stress

Nadia ERRAFIY
Adnane MOUTAOUAKKIL

Oxidative and nitrosative stress

Cellular response and natural remedy

ScienciaScripts

Imprint

Any brand names and product names mentioned in this book are subject to trademark, brand or patent protection and are trademarks or registered trademarks of their respective holders. The use of brand names, product names, common names, trade names, product descriptions etc. even without a particular marking in this work is in no way to be construed to mean that such names may be regarded as unrestricted in respect of trademark and brand protection legislation and could thus be used by anyone.

Cover image: www.ingimage.com

This book is a translation from the original published under ISBN 978-3-8416-3613-3.

Publisher:
Sciencia Scripts
is a trademark of
Dodo Books Indian Ocean Ltd. and OmniScriptum S.R.L publishing group

120 High Road, East Finchley, London, N2 9ED, United Kingdom
Str. Armeneasca 28/1, office 1, Chisinau MD-2012, Republic of Moldova, Europe
Printed at: see last page
ISBN: 978-620-7-70530-6

Contents

INTRODUCTION

In recent years, the world of biological and medical sciences has been invaded by a new concept, that of 'oxidative stress'. This is a situation in which the cell no longer controls the excessive presence of free radicals. A situation that researchers implicate in most diseases, such as cancer (Gerber et *al.*, 2002), atherosclerosis, neurodegenerative diseases (Kohen and Nyska, 2002), rheumatoid arthritis, ischemia/reperfusion and the degenerative process of biological ageing (Harman, 1956; Harman, 1981; Gutteridge and Halliwel, 2000).

Oxidative stress is defined as an imbalance between antioxidant and oxidant systems in favour of the latter, leading to the production of reactive oxygen species (ROS), a source of potential toxic effects. However, the formation of reactive species is not always accompanied by toxicity. In particular, some reactive species are intermediates in normal physiological processes (cellular respiration or inflammatory phenomena). Thus, under normal physiological conditions, free radicals are formed in the body as by-products of certain biological phenomena. However, rigorous control of their formation and elimination by means of various antioxidant systems protects cells from their deleterious effects. Overproduction of oxidants and/or malfunctioning antioxidants can upset the oxidative balance, leading to cellular stress. To avoid the consequences of oxidative stress, it is necessary to re-establish the oxidant/antioxidant balance in order to preserve the body's physiological performance.

In this context, the use of natural antioxidants is an area of growing interest. Some of these can be found in vegetables, fruit, spices and plant extracts. Aromatic and medicinal plants (MAPs) are also an inexhaustible source of substances with a wide range of biological and pharmacological activities. Essential oils are secondary metabolites extracted from plants known for their therapeutic properties.

On the basis of this information and much more, we set out to study and evaluate the stress caused by two stress agents on a unicellular eukaryotic microorganism known for its various applications as a cellular model in the field of scientific research: *Tetrahymena thermophila*. This protozoan has all the systems found in a human cell: nucleus, mitochondria, cytoskeleton and systems for secreting and responding to chemical messengers, which is why it is being considered as an alternative model to animal cells, particularly mammalian cells, for this type of study. Since oxidative stress can be minimised by the addition of antioxidants, we were also interested in studying the antioxidant potential of essential oils on protozoa subjected to oxidative stress.

Thus, the work présentë in this book focuses on three main areas:

The first axis focused on the study of the effects caused by the two stress agents (H_2O_2 and SNP) on the protozoan *T. thermophila* chosen as the cellular modèle. To this end, we evaluated the effect of oxidative stress caused by hydrogen peroxide (H_2O_2) and nitrosative stress caused by sodium nitroprusside (SNP) on the growth, morphology and physiology of the protozoan.

In the second axis, the key enzyme of carbohydrate metabolism, glyceraldehyde-3-phosphate dehydrogenase (GAPDH), was chosen as an enzymatic model to assess the physiological perturbations caused by stress in the protozoan *T. thermophila*. We therefore focused on the purification and characterisation of this enzyme in *T. thermophila*. We therefore set up an experimental purification protocol using fractional precipitation with ammonium sulphate followed by two column chromatographies (DEAE-cellulose and Mono-S). This protocol enabled us not only to obtain GAPDH from *T. thermophila* with a high degree of purity, but also to characterise it for the first time to our knowledge. The pure enzyme was used as a metabolic marker to assess the *in vitro* effects of H2O2-induced oxidative stress and SNP-induced nitrosative stress.

Finally, in the third axis, we studied the antioxidant potential of essential oils from certain aromatic and medicinal plants on *T. thermophila* exposed to oxidative or nitrosative stress in order to verify whether a supply of essential oils can reduce the effects of these stresses to which the protozoan is subjected. The synergistic interaction of essential oils was also evaluated in order to improve the antioxidant potential of *T. thermophila* exposed to stress.

CHAPTER 1

BIBLIOGRAPHICAL STUDY

I. STRESS

I. 1 History of stress

The concept of stress has existed since the Middle Ages. The first studies on stress date back to the end of the XIXeme century with the French physiologist Claude Bernard, who introduced the notion of the constancy of the internal environment, the principle of which is based on the fact that all living beings must maintain a certain internal stability in the face of continuous changes in the external environment. Fifty years later, the idea was developed by the American physiologist Walter B. Cannon in his book 'The Worm'. Cannon in his book *The Wisdom of the Body* (Cannon, 1932). In it, he describes the mechanisms governing this bodily constancy, which he calls homëostasia. He was the first to use the word stress (from the Latin *stringer*, meaning to make stiff, to squeeze), which he borrowed from the vocabulary of mechanics to describe the stresses likely to disrupt homëostasis. However, the real turning point came in the 1940s-1950s, with the work of the Hungarian-born Canadian endocrinologist Hans Selye, Director of the Institute of Experimental Medicine and Surgery at the University of Montreal, who developed the first complete theory of 'medical' stress by defining stress as a non-specific response given by the body to any demand made upon it. He called this response the "general adaptation syndrome" and distinguished three phases: alarm, resistance and exhaustion. The first corresponds to all the body's responses to a sudden disturbance, the second to those put in place should the disturbance persist. The exhaustion phase occurs when the body is no longer able to adapt. This is followed by the many complications of stress, often characterised by inflammatory diseases.

I. 2. Oxidative/nitrosative stress

I. 2. 1 Definition

Oxidative/nitrosative stress is defined as an imbalance in the cellular metabolic balance during which the generation of oxidants overwhelms the antioxidant defence system (Figure 1), resulting in the accumulation of free radicals causing oxidative damage to macromolecules (Repine et *al.*, 1997). Free radicals involving one or more oxygen atoms are called reactive oxygen species (ROS) and those involving one or more nitrogen atoms are called reactive nitrogen species (RNS). These free radicals are chemical species, atomic or molecular, containing one or more unpaired free electron(s) on their outer layers (Lehucher-Michel et *al.*, 2001). Reactive species include molecules such as superoxide anion ($O_2^{\cdot-}$), hydroxyl radical ($OH^{\cdot}$) (Edreva, 2005), hydrogen peroxide (H_2O_2) (Briviba et *al.*, 1997), nitric oxide ($NO^{\cdot}$) and peroxynitrite ($ONOO^{-}$) (Beckman, 1996). Although free radicals are generated during normal biological processes as by-products of metabolism (Figure 2) and play an important role in certain biological functions, such as phagocytosis, regulation of cell growth and intercellular signalling, their ability to modify various molecules in the body is not sufficient to prevent the formation of free radicals, Their ability to modify various molecules in a harmful way is blocked by a variety of intra- and extracellular antioxidant systems, including enzymatic systems (catalase, superoxide dismutase (SOD), glutathione peroxidase) and non-enzymatic systems (glutathione, vitamin C, vitamin E). A deficit in these defences leads to an increase in oxidative stress (Halliwell, 1992) and in the event of an excess of ROS/RNS, the latter can react with fatty acids, proteins and DNA, causing damage to these substrates. Cytotoxicity has been derived from the imbalance between ROS/RNS production and intracellular antioxidant defence mechanisms (Repine et *al.*, 1997). All this leads to major changes in the structure-function relationship in any organ, system or group of specialised cells in the body. Consequently, oxidative/nitrosative stress is recognised as a general mechanism of cellular damage implicated in the pathogenesis of many diseases (Ames et *al.*, 1993).

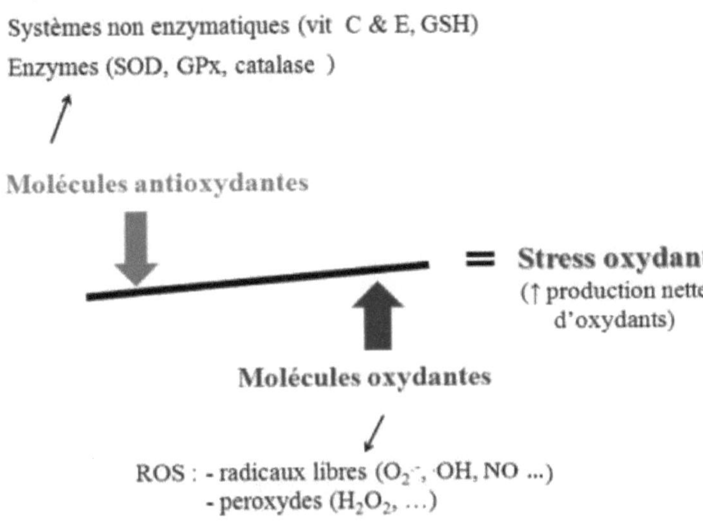

Figure 1: Balance between oxidising and antioxidising molecules

GPx: glutathione peroxidase, GSH: glutathione reductase, H2O2: hydrogen peroxide, Q?':
superoxide
anion, OH⁻ : hydroxyl radical, NO: nitric oxide, SOD: superoxide dismutase

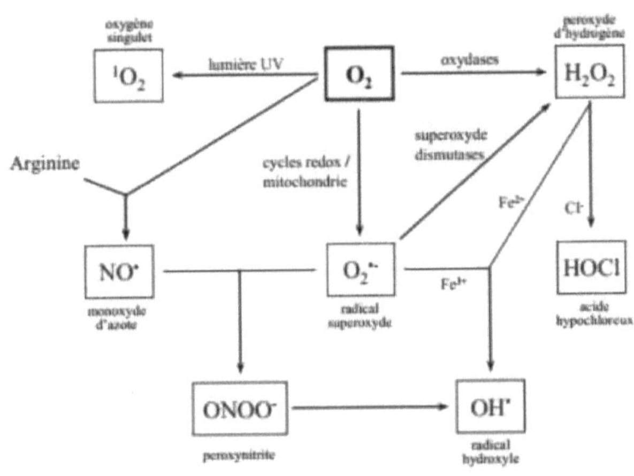

Figure 2: Mechanism of endogenous production of ROS and RNS

1.   2.2 Sources of ROS

Several endogenous and exogenous systems can produce reactive oxygen species (Stocker and Keaney, 2004).

1. 2. 2. a- Endogenous sources

The main endogenous sources of ROS are mitochondria, peroxisomes, cytochrome p450 and NADPH oxidase (Figure 3).

- The mitochondria

The mitochondrion, an organelle that uses oxygen to produce ATP, is the main source of free radicals in eukaryotic cells (Yu, 1994). Oxygen-reactive species are mainly produced during oxidative phosphorylation; a chain of enzymatically catalysed coupled reactions by a sërie of situc complexes in the inner membrane. These complexes are numbered from I to IV. Free radicals are produced mainly in complexes I and III of the clectron transport chain (Cross et jones, 1991; Finaud et *al.*, 2006). Approximately 5-10% of the total oxygen reaching the mitochondria is reduced by the action of the clectrons from the respiratory chain transporters to superoxide anion ($o_2^{\cdot-}$) which, through the action of superoxide dismutase, gives hydrogen peroxide, which in turn becomes the hydroxyl radical through the Haber-Weiss reaction (Boveris and Cadenas, 1975).

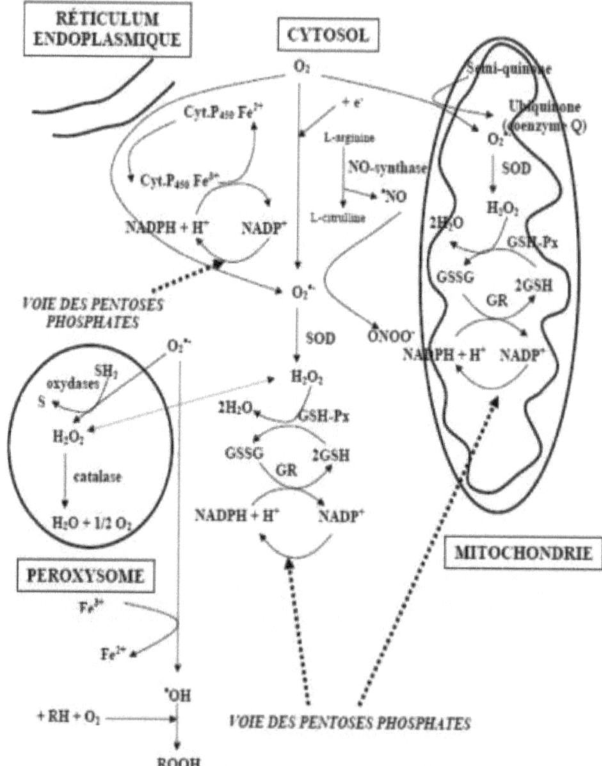

Figure 3: Main cellular sources of ROS/RNS (*afterBonnefont-Rousselot, 2003*)

- Peroxisomes

Peroxisomes are intracellular organelles bounded by a single membrane and dëpourvous of DNA. They participate in numerous metabolic pathways such as the ß- oxidation of very long-chain fatty acids. These organelles have a high capacity to produce hydrogen peroxide (Boveris and Chance, 1973) due to the high content of oxidases. These oxidases catalyse the reduction of divalent oxygen without the formation of superoxide radicals.

- **Cytochrome p450**

This system is capable of reducing substances with the subsequent formation of free radicals by a monovalent process. Several studies have shown the involvement of this system in the production of free radicals. For example, the elimination of xenobiotics and strange chemical compounds by the action of cytochrome p450 leads to the formation of free radicals. Indeed, it has been shown that in type II alveolar monocytes, the reduction of Paraquat (a powerful herbicide) leads to the formation of a cationic radical which reacts with molecular oxygen to produce O_2^{-} (Ryrfeldt et *al.*, 1993). Another study also showed that in the liver, elimination of ethanol by cytochrome p450 leads to the formation of the hydroxyethyl radical (CH3-C·HOH) (Poli, 1993).

- **NADPH oxidase**

NADPH oxidase is considered to be the most important source of superoxide anion in vascular cells and phagocytes (Griendling et *al.*, 2000). NADPH oxidase uses oxygen to produce large quantities of superoxide anion in the cell membrane. In macrophages, NADPH oxidase eliminates sequestered bacteria by producing ROS. It has been shown that H2O2 can activate NADPH oxidases in non-phagocytic cells to induce an increase in ROS concentration (Li et *al.*, 2001).

I. 2. 2. b- Exogenous sources

In addition to endogenous systems, there are exogenous sources that produce ROS. The environment around us encourages exposure to the many free radicals generated by pollution, sunlight and other radiation. Substances such as nitrogen dioxide, ozone and nitric oxide can be found in the air. Tobacco smoke, in turn, contains high concentrations of nitric oxide, peroxide radicals and carbon, making it a major source of free radicals. Similarly, alcohol and certain drugs are major sources of free radicals generated by their oxidation at cytochrome p450 (Favier, 2003). On the other hand, both ionising radiation and solar rays can activate a large number of atoms and molecules, particularly oxygen, favouring the production of excited triplets by changing electron orbitals (Foote, 1982).

I. 2. 3 Sources of RNS

Nitric oxide (NO), produced under physiological conditions, regulates countless cellular functions by post-translational modification of various proteins. This occurs mainly at the level of cysteine residues (S-nitrosylation). At physiological concentrations, superoxide anion promotes this reaction. However, at pathological concentrations, this molecule interferes with S-nitrosylation, by interacting directly with proteins and joining with nitric oxide, thus forming RNS. Peroxynitrite (ONOO⁻), like other RNS, is formed by the reaction of superoxide anion with nitric oxide (Beckman et *al.*, 1990).

I. 2. 4 Types of ROS/RNS

There are various types of free radical derived from oxygen and nitrogen. The most important are summarised below:

I. 2. 4. a- Superoxide anion (O_2^{-})

The superoxide radical arises from the monovalent reduction of molecular oxygen, which captures an electron. Unlike other free radicals characterised by their high reactivity, the superoxide anion is only capable of reacting effectively with a limited number of molecules (Halliwell, 1996). The dismutation of this superoxide anion leads to the formation of fundamental oxygen and hydrogen peroxide.

$$2\,O_2^{\cdot-} + 2H^+ \rightarrow H_2O_2 + O_2$$

On the other hand, the superoxide anion can act as a Bronsted base, capturing a proton and giving rise to the hydroperoxyl radical ($HO_2^{\cdot}$) which is more reactive than $O_2^{\cdot-}$ (Fridovich, 1997).

$$O_2^{\cdot-} + H^+ \rightarrow HO_2^{\cdot}$$

I. 2. 4. b- Hydrogen peroxide (H_2O_2)

Hydrogen peroxide, having no unpaired electrons on its outer layer, is not a free radical in the true sense of the word. It is the least reactive form of ROS. Its importance lies in the fact that it participates in numerous reactions leading to the generation of free radicals. In addition, its ability to cross biological membranes means that it can trigger oxidation reactions in the cell at a great distance from where it is produced.

It can come from a variety of sources:

*   Direct reduction of an oxygen molecule by two electrons (Fridovich, 1997):

$$2O_2 + 2e^- + 2H^+ \rightarrow H_2O_2$$

*   Dismutation of superoxide anion (Cheeseman and Slater, 1993; Frei, 1994),
*   Product of certain enzymes such as glucose oxidase, which catalyses the oxidation of glucose to hydrogen peroxide and D-glucono-6-lactone (Fridovich, 1986),
*   Rë chemical reactions such as the copper-catalysed auto-oxidation of ascorbic acid (Korycka-Dahi and Richardson, 1981).

1.   2. 4. c- The hydroxyl radical (OH)

It is obtained when molecular oxygen is reduced by 3 electrons. It is the most unstable and reactive of all the oxygen dërivëes. It has a half-life in the nanosecond range, allowing it to react with numerous molecular species (proteins, lipids, DNA) in the vicinity of its formation site (Cheeseman and Slater, 1993). This radical is generated by different processes:

- Lysis of the water molecule or hydrogen peroxide by the action of ionising radiation (Cheeseman and Slater, 1993). The hydroxyl radical can be formed *in vivo* following high-energy radiation (X-rays, Y-RAYS) which causes homolytic disruption of the water molecule (a). UV light does not have sufficient energy to split a water molecule, but it can split hydrogen peroxide into two hydroxyl radical molecules (b).

(a) $H_2O + h\upsilon \rightarrow OH^{\cdot} + H^{\cdot}$

(b) $H_2O_2 + UV \rightarrow 2OH^{\cdot}$

*   Reduction of H2O2 by transition metals such as iron or copper (Frei, 1994). The most important process in the formation of the hydroxyl radical is the Fenton reaction.

$$H_2O_2 + Fe^{2+} \rightarrow Fe^{3+} + OH^- + OH^{\cdot}$$

*   It can also be generated from hydrogen peroxide and the superoxide radical by the Haber-Weiss reaction.

$$H_2O_2 + O_2^{\cdot-} \rightarrow O_2 + OH^- + OH^{\cdot}$$

1.   2. 4. d- Nitric oxide (NO)

Nitric oxide (NO) belongs to the group of reactive nitrogen species (RNS). It is an inorganic gas with a free electron, which gives it great chemical instability, a characteristic common to all free radicals. It is a low molecular weight molecule that is highly lipophilic and water-soluble, with a relatively long half-life (3-5 seconds) compared with other ROS. In its radical form (NO$^{\cdot}$), nitric oxide has an unpaired electron in its outer orbital. The loss of this electron results in the formation of the nitrosonium cation (NO^{+}), while the gain of an electron will form the nitroxyl radical (NO^{-}). Each of these compounds has specific properties and reactivates (Stamler et *al.*, 1992). Via its various redox forms, NO constitutes a precursor of other more reactive species by reacting with O2^{-} to form a powerful oxidant peroxynitrite (ONOO^), which can secondarily decompose into other oxidants such as NO2 and OH (Densiov and Afanas'ev, 2005). It is formed by an enzymatic reaction in which nitric oxide synthase (NOS) catalyses the conversion of L-arginine to L-citrulline (Figure 4) (Moncada et *al.*, 1991). It is synthesised by a wide variety of cell types, including macrophages, endothelial cells, neutrophils, hepatocytes and neurons. It plays a fundamental role in many physiological processes, in particular the regulation of blood pressure, the defence mechanism, inhibition of platelet aggregation, neurotransmission and immune regulation (Valko et *al.*, 2007). Studies in micro-organisms have shown that NO induces the expression of genes involved in responses to oxidative stress and signal transduction in pathogenicity and resistance (Crane et *al.*, 2010; Meilhoc et *al.*, 2010).

$$O2 + Arginine + NADPH \wedge NO. + Citrulline + H2O + NADP+.$$

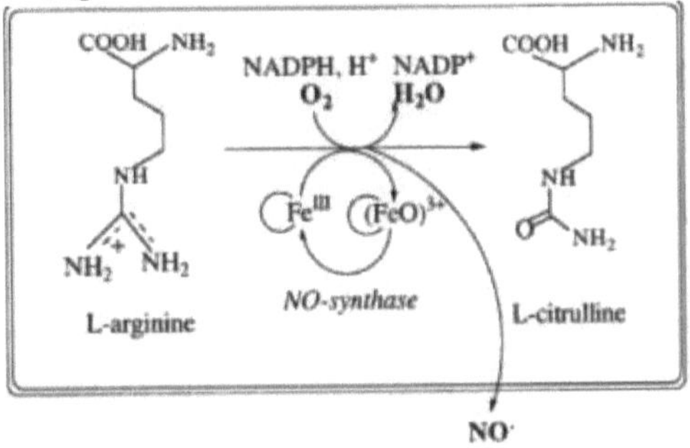

Figure 4: NO synthesis by the enzyme NO-synthase

NO and some of its derivatives are involved in modulating protein activity via three main types of post-translational modification: metal-nitrosylation, tyrosine (Tyr) nitration and S-nitrosylation (Figure 5) (Besson-bard et *al.*, 2008; Hess et *al.*, 2005).

- **Metal-nitrosylation**

Metal-nitrosylation is a reversible mechanism involving the establishment of a covalent bond between NO$^{\cdot}$ and a transition metal of a metalloprotein. NO$^{\cdot}$, as an electron donor, reacts with transition mëtals such as iron, copper or zinc, leading to the formation of mëtal-nitrosyl complexes.

- **Tyrosine nitration**

This involves the addition of an NO2+ group to a Tyr residue of a protëine, leading to the formation of 3-nitrosotyrosine. This post-translational modification is generally irreversible (Hanafy et *al.*, 2001).

- **S-nitrosylation**

S-nitrosylation involves the covalent incorporation of NO into the thiol group of a cystëine

(Cys) residue of a protëine leading to the formation of S-nitrosothiol (S-NO, Hess et *al.*, 2005). S-nitrosylation is a form of post-translational modification of protëines exhibiting similarities to phosphorylation (Anand and Stamler, 2012). It is considerëdërëe as a signalling mëcanism (Hoffmann et *al.*, 2003), prëcisëment cibM (Sun et *al.*, 2001), reversible and necessary for spëcific cellular responses (Hess et *al.*, 2005).

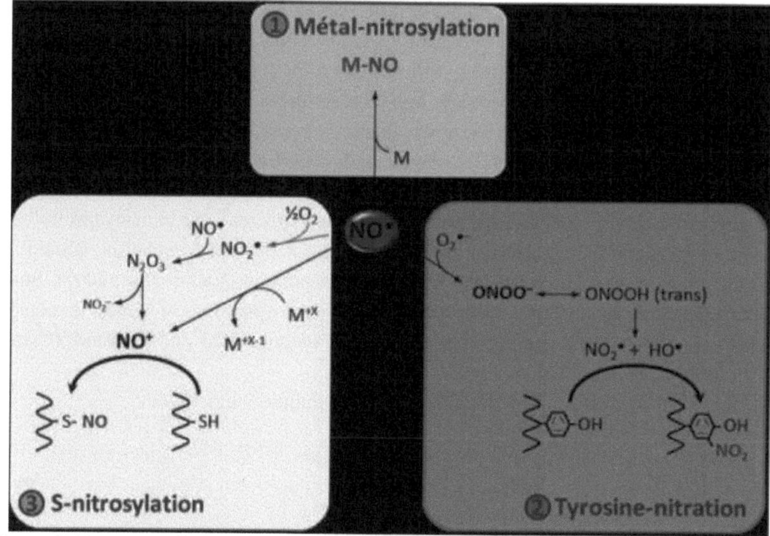

Figure 5: Post-translational modifications induced by NO (*after Besson-Bard et al., 2008*)

I. 2. 5 ROS/RNS functions

ROS and RNS are known to play a dual role; as beneficial species and as harmful species. At low physiological levels, they serve as messenger signalling mediating various biological reactions, including gene expression, cell proliferation, angiogenesis, innate immunity, programmed cell death and senescence (Dowling and Simmons, 2009; Scherz-shouval and Elazar, 2011). On the other hand, at high levels, these reactive molecules can have harmful effects by causing oxidative damage to biological macromolecules and disrupting the cellular redox balance (Dowling and Simmons, 2009; Acharya et *al.*, 2010).

Disruption of homeostasis by ROS/RNS is generally considered to be a risk factor for the initiation and progression of diseases such as atherosclerosis, diabetes, neurodegeneration and cancer (Dowling and Simmons, 2009; Salmon et *al.*, 2010). The beneficial or harmful effects of ROS/RNS depend on the site, type and quantity of ROS/RNS produced, as well as the activity of the antioxidant defence system (Circu and Aw, 2010).

Endogenous ROS/RNS can be generated as the main function of an enzyme system (e.g. NADPH oxidases that are activated in response to active receptors), as by-products of other biological reactions (the mitochondrial electron transport chain) or by metal-catalysed oxidations (Fenton reaction) (Powers and Jackson, 2008).

I. 2. 6 Biological consequences of ROS/RNS

An excess of ROS/RNS results in the appearance of often irreversible cellular damage, with the most vulnerable biological targets being DNA, lipids and proteins (Halliwell and Whiteman, 2004; Valko et *al.*, 2006) (Figure 6).

I. 2. 6. a- Oxidation of DNA

DNA is highly sensitive to free radical attack. Purine and pyrimidine bases and desoxyribose are the prime targets of ROS/RNS. They are then transformed into fragmentation products and oxidised bases (Martinez-Cayuela, 1995). Free radicals, particularly the hydroxyl radical, have a strong affinity to react with purine and pyrimidine bases as well as with pentose riboses and desoxyriboses, thereby modifying the structure of DNA. ROS/RNS induce an increase in the number of mutations, overlaps, chromatid breaks and the loss of chromosomal fragments (Roche and Romero-Alvira, 1996).

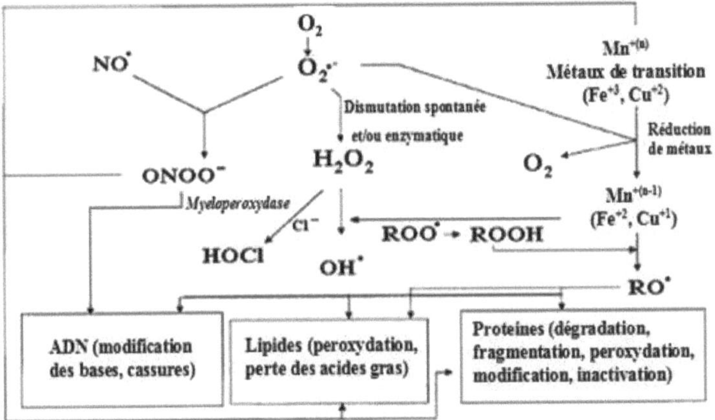

Figure 6: Biological targets of ROS/RNS (*after Kohen and Nyska, 2002*)

The interaction of ROS with DNA can occur directly (in the case of OH·) or indirectly in the presence of transition metals that allow the subsequent formation of hydroxyl radicals (in the case of H2O2 and O2⁻). It should be noted in this respect that H2O2 is highly toxic due to its ability to cross membranes and produce hydroxyl radicals.

I. 2. 6. b- Oxidation of proteins

In proteins, amino acids are a target of ROS/RNS, either at the side chain with the formation of oxidation products, or at the peptide bond leading to fragmentation of the chain (Berlett and Stadtman, 1997). Radical damage to proteins can be broadly classified into two types: (i) diffuse attacks causing general changes and (ii) targeted attacks causing changes at specific points on the protein. The former (i) are the breaks observed following exposure to ionising radiation and ozone. The second (ii) are caused by oxidised derivatives of the lipid or carbohydrate type that react with the functional groups of proteins (Stadtman and Berlett, 1998). The most sensitive amino acids are sulphur amino acids (cysteine and methionine) and aromatic amino acids (tyrosine and tryptophan). Amino acid oxidation generates hydroxyl and carbonyl groups on proteins but can also induce more significant structural modifications such as intra- or intermolecular cross-linking, which affects their function, antigenicity and activity (Martinez-Cayuela, 1995; Lehucher-Michel et *al.*, 2001; Valko et *al.*, 2007). Modified proteins generally become more sensitive to the action of proteases and are then directed towards proteolytic degradation by the proteasome (Jung et *al.*, 2007).

I. 2. 6. c- Lipid peroxidation

Polyunsaturated fatty acids (PUFAs) are prime targets for ROS/RNS (Davies, 2000), whose oxidative damage is known as lipid peroxidation. Lipid peroxidation can be defined as oxidative damage to polyunsaturated fatty acids produced by an uncontrollable autocatalytic process (Figure 7). It represents a form of tissue and cell damage production usually initiated by ROS/RNS, forming a cascade of reactions with the production of free radicals and the formation of organic peroxides and other products from unsaturated fatty acids (Dias et al., 1998). The oxidation products formed can participate as second messengers in the regulation of metabolic functions, gene expression and cell proliferation. Above all, they can also act as toxic substances responsible for cell dysfunction and damage (loss of polyunsaturated fatty acids, reduced membrane fluidity, altered activity of enzymes and membrane receptors, release of material from the subcellular compartment) (Beckman and Ames, 1998; Lehucher-Michel, 2001).

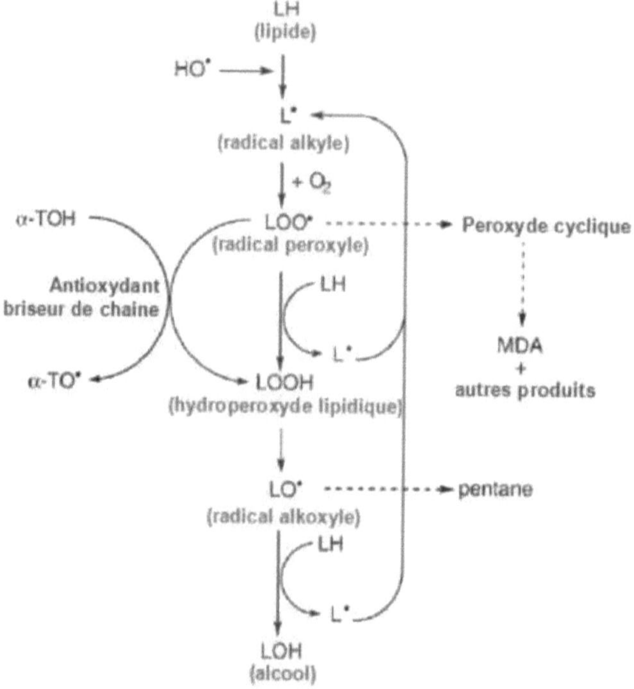

Figure 7: Lipid peroxidation (*after Sachdev and Davies, 2008*)

HO\ Hydroxyl radical, LH: Fatty acid, MDA: Malondialdehyde,
O2 : Oxygene, a-TOH: Tocopherol, a-TO : Derive tocopheroxyl

Lipid peroxidation is a chain reaction which takes place in three phases (Figure 7):

- **Introduction :**

This ëtape begins with the attack of unsaturated fatty acids by free radicals, with the subtraction of a hydrogen atom from a methyl of the lipid chain, generating a carbon radical (Halliwell and Chirico, 1993). Among the free radicals that cause these processes are the hydroxyl radical and the protonated form of the superoxide anion (Bielski et al., 1983) (OH• removes a hydrogen atom from CH2 and then the double bonds undergo molecular rearrangement leading to the formation of conjugated dienes), in the presence of O2 the carbon radical is transformed into the peroxyl radical (RO· 2) (Martinez-Cayuela, 1995).

- **Propagation :**

Also known as secondary initiation. This is the activation of peroxidation by a lipid radical. The RO· 2 radical removes a hydrogen from a new neighbouring PUFA, which in turn produces an R· radical and then an RO· 2 radical. An autocatalytic chain reaction sets in. In the presence of transition metals, the hydroperoxides (LOOH) formed can undergo cleavage at C-C bonds to give rise to various decomposition products, notably malondialdehyde (MDA) and 4-hydroxynoneal, the most toxic products of lipid peroxidation (Martinez-Cayuela, 1995; Lehucher-Michel, 2001).

- **Termination :**

This phase consists of forming stable (non-radical) compounds resulting from the meeting of two radical species or, more often, from the reaction of a radical with an antioxidant molecule known as a "chain breaker" (Kohen and Nyska, 2002).

II. ANTIOXIDANTS

II. 1 Definition

To counterbalance oxidative and nitrosative stress, cells use a wide range of enzymatic and non-enzymatic defence mechanisms known as antioxidants (Powers and Jackson, 2008). An antioxidant can be defined as a substance capable, at low concentration, of entering into competition with other oxidisable substrates, and thus slowing down or inhibiting the oxidation of these substrates (Berger, 2006).

II. 2 Categories of antioxidants

II. 2. 1 Enzymatic antioxidants

Enzymatic antioxidants have a high affinity for ROS. They react very quickly with these species to neutralise them (Muzykantov, 2001). The main enzymes involved are catalase (Cat), superoxide dismutase (SOD), glutathione peroxidase (GPx) and glutathione reductase (GR) (Figure 8).

II. 2. 1. a- Superoxide dismutase

SOD is one of the first antioxidant defence barriers, catalysing the dismutation of $O2^-$ into $H2O2$ in the following reaction: $2O2^- + 2H^+ \wedge H2O2 + o2$. There are three isoforms of SOD: ferrous SOD (Fe-SOD), copper SOD (Cu-SOD) and manganese SOD (Mn-SOD), which differ according to the chromosomal location of the gene, their metal content, their quaternary structure and their cellular location (Zelko et al., 2002).

II. 2. 1. b- Catalase

Catalase is an enzyme that catalyses the reaction of detoxification of $H2O2$ by transformation into $H2O$ and O2 according to the following reaction:

$$2\,H_2O_2 \rightarrow O_2 + 2\,H_2O$$ (Wassman et al., 2004)

Catalase is normally found in peroxisomes and is most active when oxidative stress levels are high. It plays a significant role in the development of tolerance to oxidative stress in the adaptive response of cells (Wassman et al., 2004).

II. 2. 1. c- Glutathione peroxidase and glutathione reductase

GPx and GR are located in the cytosol and in the mitochondria. GPx is a protein made up of 4 identical subunits, each containing a selenocysteine residue essential for its enzymatic activity. It plays an important role in $H2O2$ detoxification. It acts in synergy with SOD, since its role is to accelerate the dismutation of $H2O2$ into $H2O$ and o2, but it can also catalyse the reduction of other hydroperoxides (ROOH) using the reducing power of glutathione (GSH).

The role of GR is to regenerate GSH from glutathione disulphide (GSSG) using NADPH as a cofactor (Martinez-cayuela, 1995; Sorg, 2004).

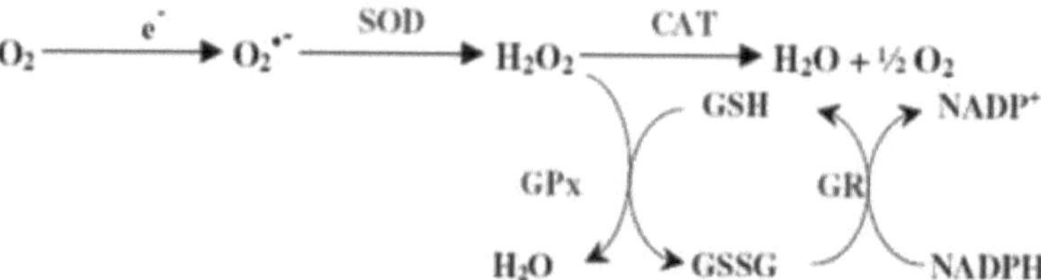

Figure 8: Enzymatic antioxidants

II. 2. 2 Non-enzymatic antioxidants

This group of antioxidants includes endogenous substances synthëtiséed by the cell (glutathione) and exogenous substances supplied by the diet, such as vitamin E (a-tocopherol) and vitamin C (ascorbic acid). The latter act by trapping free radicals by neutralising unpaired electrons, allowing them to be transformed into stable molecules (Pincemail et *al.*, 2002; Koechlin-ramonatxo, 2006).

II. 2. 2. a- Glutathione

Reduced glutathione (GSH) enables H2O2 to be reduced to H2O via the reaction catalysed by glutathione peroxidase (GPx). It can also reduce the radicals formed by the oxidation of vitamins E and C, thereby lowering the level of lipid peroxidation (Packer et *al.*, 1997; Powers and Lennon, 1999). The ratio of reduced glutathione to glutathione oxide (GSH/GSSG) is often used as a marker of oxidative stress because the greater the H2O2 flux, the more reduced glutathione is consumed and the higher the glutathione oxide (Ji and Mitchell, 1992).

II. 2. 2. b- Vitamin E

Because of its liposolubility, vitamin E (a-tocopherol) is the most important antioxidant in lipids (Vertuani et *al.*, 2004). It binds to cell membranes, thereby sequestering free radicals and preventing the propagation of lipid peroxidation reactions (Packer et *al.*, 1997; Evans, 2000). Vitamin E acts in two different ways, either by trapping ROS directly or by up-regulating antioxidant enzymes.

II. 2. 2. c- Vitamin C

Vitamin C, being water-soluble, is essentially found in the cytosol and extracellular fluid. Its action can be direct, by capturing O2" and OH, or indirect, by contributing to the regeneration of a-tocopherol, thus making vitamin E more effective (Packer et *al.*, 1997; Evans, 2000).

On the other hand, as the role of antioxidants has been established in stimulating cellular defences against oxidative stress, an exogenous supply of antioxidants can contribute effectively to counteracting the harmful effects of this stress. Most of these exogenous antioxidants come from plants (fruit, vegetables or plants). Thus, essential oils and secondary plant metabolites could play a major role in stress prevention and/or remëdiation.

II.3 Essential oils

II. 3. 1 General points and definition

Essential oils, also called essences, are oily, volatile and odorifërant products formedë by aromatic plants as secondary mëtabolites and present, in small quantities relative to the mass of the vëgëtal, in the form of tiny droplets in leaves, fruit skin, resin, branches, roots and wood bark. They can be obtained by expression, fermentation, enfleurage or extraction. The most commonly used method is steam distillation (hydrodistillation). Essential oils are known for their antiseptic properties. In addition to their antibacterial properties (Deans and Ritchie, 1987; Mourey and Canillac, 2002), essential oils or their components have been shown to have antiviral (Koch et *al.*, 2008), antifungal (Zabka et *al.*, 2009; Mishra et *al.*, 2012), anti-toxigenic (Atanda et *al.*, 2007), anti-parasitic (Machado et *al.*, 2011) and insecticidal (Maciel et *al.*, 2010; Tozlua et *al.*, 2011) properties. Some essential oils appear to have specific medicinal properties for healing organ dysfunction or systemic disorders (Silva et *al.*, 2003; Perry et *al.*, 2003).

Around 3,000 essential oils are currently known worldwide. Three hundred of these are marketed, mainly for the pharmaceutical, agronomic, food, health, cosmetics and perfume industries.

II.3.2 Natural state and distribution

Essential oils are produced within the cytoplasm of certain plant cells. They separate from the cytoplasm by syneresis in the form of small droplets which then confluence to form more or less extensive patches. These secretory cells can be found in all the plant's organs (vegetative organs and reproductive organs), such as secretory hairs or essence cells. Essential oils are liquid, volatile, clear, rarely coloured, soluble in lipids and in organic solvents with a density generally lower than that of water.

In nature, essential oils play an important role in protecting plants, as antibacterial, antiviral, antifungal, insecticide and herbicide. They can also attract certain insects to help disperse pollen and seeds, or repel other undesirables.

Essential oils are extracted from a variety of aromatic plants, generally found in temperate regions of the world, but also in hot climates such as the Mediterranean and tropical countries, where they form an important part of traditional pharmacopoeia.

II. 3.3 Chemical composition

Essential oils are complex mixtures that may contain more than 60% of different compounds. They are characterised by two or three major components in fairly high concentrations (20-70%) compared with other components present in minute quantities (Senatore, 1996). Generally speaking, these major components determine the biological properties of essential oils. These constituents are mainly terpene carbides and their oxidation derivatives, as well as aromatic and aliphatic derivatives, characterised by low molecular weights (Bakkali et *al.*, 2007).

II. 3. 4 Classification of essential oils

11. 3. 4. a- Terpenes

Terpenes are natural hydrocarbons with a cyclic or linear structure. Structurally and functionally, they form different classes. The union of two molecules with 5 or more carbon atoms (isoprene) allows the majority of terpenes to be constructed theoretically. This is Ruzicka's (1953) isoprenic chain rule. However, isoprene has never been found free in nature. Their classification is based on the number of repetitions of the basic isoprene unit: hemiterpenes (C5), monoterpenes (C10), sesquiterpenes (C15), diterpenes (C20), sesterpenes (C25), triterpenes (C30), tetraterpenes (C40) and polyterpenes. A terpene containing an oxygen is called a terpenoid. The reactivity of the intermediate cations obtained during the biosynthetic process of mono- and sesquiterpenes explains the existence of a large number of derived molecules.

- **Hemiterpenes**: there are few natural compounds with a branched C5 formula. Only isoprene has all the biogenetic characteristics of terpenes (Loomis and Croteau, 1980).

- **Monoterpenes**: derived from the coupling of two isoprene (C10) units (Allen et *al.*, 1977). They are the most representative molecules, making up 90% of essential oils and allowing a wide variety of structures (acyclic, monocyclic and dicyclic). They are made up of several functions, such as carbides, alcohols, aldehydes, cetones, esters, ethers, phenols and peroxides.

- **Sesquiterpenes**: are formed by the assembly of three isoprene units (C15). Extending the chain increases the number of cyclisations, allowing a wide variety of structures. The structure and function of sesquiterpenes are similar to those of monoterpenes.

II. 3. 4. b- Aromatic compounds

Derived from phënylpropane, aromatic compounds are less common than terpenes. The pathways for the biosynthesis of terpenes and phenylpropane derivatives are generally separate in plants but can co-exist in some cases. These aromatic compounds include aldehydes (cinnamaldehyde), alcohols (cinnamic alcohol), phenols (chavicol and eugenol), methoxy derivatives (methyleugenol) and methylenedioxy compounds (safrole). Plants containing large quantities of these compounds are aniseed, cinnamon, cloves, fennel, tarragon, parsley and certain plants belonging to the following botanical families: Apiaceae, Lamiaceae, Myrtaceae and Rutaceae (Bakkali et *al.*, 2007).

II. 3. 5 Extraction of essential oils

Several extraction processes are commonly used to obtain essential oils. The choice of process varies according to the nature of the plant material to be treated (Guenther, 1972).

11. 3. 5. a- Extraction by distillation

With the exception of hesperidic essential oils (lemon, orange, etc.), most essential oils are obtained by steam distillation. This method is based on the existence of an azeotrope with a boiling point lower than the boiling points of the two compounds taken separately (the essential oil and the water). In this way, the volatile compounds and the water distil simultaneously at a temperature below 100°C under normal atmospheric pressure. The aromatic products are carried away by the steam without undergoing any major alterations (Franchomme and Penoël, 1990). There are three different physico-chemical processes using this principle.

- **Hydrodistillation**: This is the simplest method, and therefore the oldest in use. It consists of immersing the plant material directly in a still filled with water, placed on a heat source and then brought to the boil. The heterogeneous vapours are condensed on a cold surface and the essential oil separates by density difference (Clergeaud, 2000).

- **Steam distillation**: The plant material is not macerated directly in water. It is placed on a perforated grid above the base of the still, through which steam passes. As the steam passes through the material, the cells burst, releasing the essential oil, which is vaporised under the action of heat to form a "water-essential oil" mixture. The mixture is then transported to the condenser before being separated in the essencier into an aqueous phase and an organic phase (essential oil). This method improves the quality of the essential oil by minimising hydrolytic alterations.

- **Hydrodiffusion**: This is a variant of steam extraction. The principle of extraction is based on the downward action of a flow of steam which passes through the vëgëtal at reduced pressure. This procëdë saves extraction time and energy.

II. 3. 5. b- Extraction by expression

It is used particularly in the case of certain citrus essences (lemon and orange). The process of cold expression of the zest is carried out either manually or using a machine. The zests are dilacëresed and the contents of the ruptured sëcrëtrices pockets are rëcupërë by a physical procëdë. The classic procëdë consists of exerting, under a stream of water, an abrasive action on the surface of the fruit. Once the solid waste has been removed, the essential oil is separated from the aqueous phase by centrifugation. Other ëequipment breaks up the pockets by dëpression and collects the essential oil directly, which ëavoids dëgradations due to the action of water. In fact, most installations allow for the simultaneous or sequential recovery of fruit juice and essential oil, the latter being collected by water jet after abrasion (scratches, tips) before or during the expression of the fruit juice. Enzymatic treatment of the waste water can enable it to be recycled and significantly increases the final yield of essential oil. Citrus essential oils are also obtained directly from fruit juices by vacuum de-oiling (Clergeaud, 2000).

II. 3. 5. c- Solvent extraction

This technique involves subjecting the plant material to the repeated action of a non-aqueous solvent, which is then removed by distillation at reduced pressure to produce a semi-solid aromatic product known as concrete. The solvent may be one of the usual solvents used in organic chemistry (hexane, petroleum ether), but also fats, oils or even gases. These solvents have a higher extraction power than water, so that the extracts contain not only volatile compounds but also non-volatile compounds such as waxes, pigments, fatty acids and many others (Richard, 1992; Robert, 2000). Treatment of the concrete with ethanol allows partial separation of the fats and waxes. After distillation of the alcohol, the product obtained is called an absolute, the composition of which is similar to that of an essential oil (Proust, 2006). Extraction using organic solvents poses a problem in terms of the toxicity of the residual solvents, which is not negligible when the extract is intended for the pharmaceutical and food industries (Bruneton, 1999).

II. 3. 5. d- Extraction by enfleurage

It consists of bringing the flower into contact with a fatty substance which becomes saturated with the essence, then this fatty substance is used up by a solvent which is evaporated in a vacuum. This delicate and expensive method has been replaced by solvent extraction.

II. 3. 5. e- Extraction using supercritical fluids

This extraction technique involves the use of supercritical fluids, which have properties that differ from those of a gas or a liquid (between the two). They have a viscosity close to that of a gas and a density close to that of a liquid, with a very high diffusivity compared with a liquid fluid, which facilitates their penetration into porous media. The use of CO_2 as a supercritical fluid has become essential because it has properties intermediate between those of liquids and those of gases, giving it good extraction power. Its critical point (P = 73.8 bars, T = 31.1°C) is easily modulated by adjusting the temperature and pressure. The operating temperature enables the constituents to be extracted without denaturing them, while preserving their biological and/or organoleptic qualities. Supercritical fluid is an ideal solvent because it is natural, inert, non-flammable, non-toxic and can be easily removed from the extract without leaving residues. This technique provides extracts of very high quality (Wenqtang et al., 2007).

II. 3. 5. f- Microwave extraction

This method involves placing the plant material in a Clevenger-type device, where it is heated in a microwave oven for a short time to extract the essential oil. As a result of the heat produced by the microwaves, the sample rapidly reaches its boiling point (Kosar et al., 2005).

The volatile compounds are carried away by the water vapour formed from the plant's own water, and are then recovered using the classic processes of condensation, cooling and decantation. The composition of the essential oil obtained by this process is similar to that obtained by a steam extraction process. The small amount of water present in the system and the rapidity of the heating process means that a large proportion of the oxygenated compounds in the essential oils can be extracted, with limited thermal and hydrolytic degradation (Lucchesi et *al.*, 2007).

III. TETRAHYMENA THERMOPHILA *Tetrahymena thermophila*

Ciliated protozoa have an interesting history in the field of scientific research. Their ease of culture and their image as primitive organisms displaying the basic properties of life enabled them to be introduced into the laboratory at a very early stage. Their interest also stems from the fact that they are very different from other micro-organisms. Cilia were among the first organisms to be used to understand the genetic phenomenon of eukaryotes.

Among the ciliate protozoa, Tetrahymena is one of the most extensively studied genera. Species belonging to this genus, such as *Tetrahymena thermophila*, represent a very useful experimental system in physiology and cell and molecular biology. They combine the biological complexity of eukaryotes with the experimental accessibility of unicellular organisms.

Tetrahymena thermophila belongs to the phylum Ciliophora, to the class Oligohymenophorea (well-defined buccal cavity containing a buccal ciliate consisting of 3 adoral organelles on the left and 1 pororal organelle on the right), to the subclass Hymenostomata (ventral buccal cavity), to the order Hymenostomatida (ventral buccal ciliate) and to the suborder Tetrahymenina (Faure-Fremiet, 1956 ; Puytorac et *al.*, 1974).

III. 1 Evolutionary history of *T. thermophila*

Since 1940, ciliate protozoa have been used as model organisms for physiological and genetic research. Furgason (1940) was the first to recognise the characteristic pattern of four oral membranes in several ciliates. He named them *Tetrahymena geleii*. A decade later, Corlisse (1952) named them *Tetrahymena pyriformis*. Work on these same organisms showed that they were a group of organisms with similar morphologies but genetically different (Elliott and Gruchy, 1952). Later, with the development of research techniques, the heterogeneity of the group became evident with the identification of a large number of strains with different mating types and different biochemical and genetic characteristics. In 1976, one of these strains was given its present name of *T. thermophila* in recognition of its tolerance to high temperatures (Nanney and McCoy, 1976).

III. 2. Overview of the *T. thermophila* cell

T. thermophila is a common freshwater ciliated protozoan approximately 50 µm long and 20 µm in diameter (Figure 9). It is characterised by a complex structural order at the cell surface (Allen, 1967; Satir and Wissig, 1982). Beneath the plasma membrane, 8 distinct structural systems are observed in a precise order. From the outside, these systems are: (1) a flattened membrane containing cortical alveoli characteristic of the evolving line of alveoli. (2) 18 to 21 groups of longitudinal microtubules. (3) the membrane skeleton (epiplasm) beneath the cortical alveoli and longitudinal microtubules. (4) the ciliated units. (5) a set of secretory granules (mucocysts) arranged longitudinally. (6) a set of mitochondria aligned parallel to the rows of basal corpuscles and secretory granules (Aufderheide, 1979). (7) flattened endoplasmic reticulum sheets (Satir and Wissig, 1982) and (8) a set of small Golgi elements

(Kurz and Tiedtke, 1993).

The surface of the cell is covered with 18-21 rows of cilia, which run parallel to the antero-posterior axis of the cell and have a locomotor function. At its front end, the cell has a mouth apparatus made up of 4 elements (hence the name Tetrahymena). The buccal air contains the buccal cavity or cytostome located in the proximal part of the cilia and determines its ventral face. The buccal ciliary consists of 3 short left membranae (adoral ciliary) and a wavy right membrane (paroral ciliary). The cytoproct, which has excretory functions, is located at the posterior end of the cell on the ventral surface. There are usually two contractile vacuole pores opening into the posterior part of the cell, to the right of the cytostome-cytoproct axis and to the outside of the contractile vacuole. The latter, which mainly has an osmoregulatory role, is located in the distal part of the ^Hë.

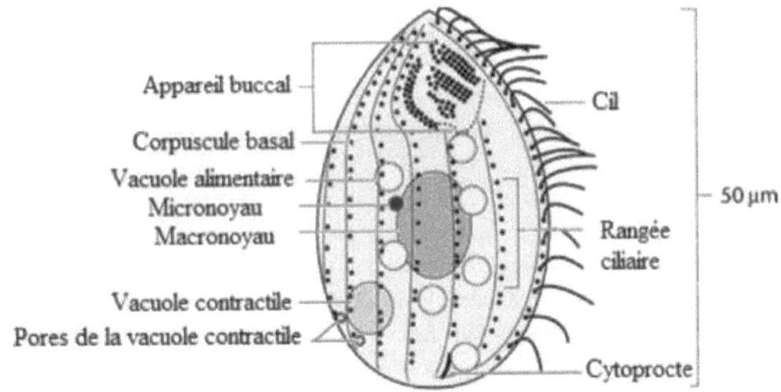

Figure 9. Diagram of the cellular organisation of *Tetrahymena thermophila*

T. thermophila is characterized by the presence in its cytoplasm of mitochondria concentrated in the cell cortex (Aufderheide, 1979), a rough endoplasmic reticulum whose relative abundance suggests a role in membrane synthesis and secreted proteins (Satir and Wissag, 1982), the Golgi apparatus associated with the endoplasmic reticulum (Franke et *al.*, 1971), the phagosome or feeding vacuole (Nilsson, 1979) and peroxisomes (DeDuve and Baudhuin, 1966).

A common feature of ciliates is the presence of nuclear dimorphism, i.e. the nuclear apparatus is composed of two structurally and functionally different types of nuclei. *T. thermophila* has a germinal micronucleus (transcriptionally inactive during vegetative growth but providing gametic nuclei through meiosis during the sexual process) and a somatic macronucleus (the seat of cellular transcription and not transmitted between sexual generations).

III. 3 Culture and growth of *T. thermophila*

T. thermophila can grow readily in a variety of axenic media with generation times of 2 to 3 h (Kiy and Tiedtke, 1992). It can reach a density of around 2 x 10^5 to 1 x 10^6 cells per millilitre.

Extensive studies on *T. thermophila* culture media have shown that this species requires 11 essential amino acids, 6 essential B complex vitamins including lipoic acid, Fe^{3+} as well as 5 other trace metals, 1 purine (guanine) and 1 pyrimidine (uracil) (Holz, 1973).

III. 4 Cell cycle of *T. thermophila*

The *T. thermophila* cell cycle comprises a variable resting period followed by significant developmental changes. These developmental changes begin with the formation of a buccal foregut, which then undergoes a complex process to form the 4 ciliary structures (membranes and undulating membrane). During the development of the buccal swelling, new ciliary units are formed inside the ciliary rows. The formation of these new units begins with the production of a new probasal organ anterior and perpendicular to the old one. The aureoles of the membrane skeleton surrounding the specialised basal bodies are formed when the new basal organs take up their positions in the cortex. An equatorial subdivision of the cellular cortex just before cytokinesis begins. An equatorial cleft appears in the ciliary rows, separating the territories of any anterior and posterior division products. Cortical organisation is asymmetrical on both sides of this cleft. A new cytoproct and new contractile vacuoles, destined for the posterior ends of the daughter cell, are formed just in front of the cleft. The most important event in the last phase of the cortical cycle is cytokinesis. The micronucleus begins to divide when the buccal membranes develop, and completes its division before the end of cytokinesis (Gavin, 1965; Lansing et *al.*, 1985). Separation of chromosome groups in *T. thermophila* coincides with subdivision of the cell zone by fission (Kaczanowska et *al.*, 1993). Micronuclear DNA synthesis in *T. thermophila* begins immediately after the completion of micronuclear division (McDonald, 1962) and ends shortly after cell sëparation. The macronuclear S phase of *T. thermophila* occurs approximately in the middle of the cell cycle and lasts about 1 h under optimal conditions (McDonald, 1962; Wolfe, 1973). This phase is probably not associated with a single phase of the cortical cycle. Macronuclear division begins at the end of micronuclear division (Jaeckel-Williams, 1978) and is thus completed by cytokinesis.

III. 5. *T. thermophila* as a model system

The common expression "model system" suggests that the system under study is either an exemplary system, or that it serves as a model for studying other systems. The conservation of many intracellular mechanisms in eukaryotes makes it possible to study a eukaryotic cellular model to shed light on the biology of human cells. Two yeasts have been widely used in this context: *Saccharomyces cerevisiae* and *Schizosaccharomyces pombe*. These cell models were chosen on the basis of their ancestral similarity to humans. However, in the course of their evolutionary history, yeasts have lost certain important ancestral structures and functions that animals and ciliates have retained, such as cilia and the regulated secretion of stored products. *T. thermophila* subsequently proved to be a suitable model for studying these intracellular systems. The excellent knowledge of the genetics of *T. thermophila,* as well as its aptitude for transformation, made this protozoan a model system of choice for the study of ancestral characteristics shared between humans and ciliates.

IV. GLYCERALDEHYDE-3-PHOSPHATE DEHYDROGENASE

Glyceraldehyde-3-phosphate dehydrogenase (GAPDH) is a ubiquitous enzyme involved in central carbon metabolism via the glycolysis pathway (Forthergill and Michels, 1993). It is known in all living organisms, including protozoa. It is one of the most evolutionarily conserved enzymes. It belongs to the family of aldehyde dehydrogenases with an NAD(P) cofactor.

IV.1 The different types of GAPDH

Glyceraldehyde-3-phosphate dehydrogenases (GAPDH, EC 1.2.1.12/13/9) are key enzymes in cellular metabolism. They are involved in both catabolic pathways, such as glycolysis or the oxidative pentose phosphate pathway, and anabolic pathways, such as gluconeogenesis or the reductive pentose phosphate pathway.

There are three classes of enzymes with glycëraldëhyde-3-phosphate dëshydrogënase activity (Martin et *al.*, 1993; Habenicht, 1997; Valverde et *al.*, 1997); these three classes differ in their enzymatic function, cellular location and cofactor spëcificitë :

- NAD'-dependent phosphorylating GAPDH (EC 1.2.1.12)
- NADP+-dependent phosphorylating GAPDH (EC1.2.1.13)
- Non-phosphorylating NADP-dependent GAPDH' -dependent (EC 1.2.1.9)

IV.1.1 NAD+-dependent phosphorylating GAPDH (EC 1.2.1.12)

The NAD-phosphorylating' -dependent GAPDH is located in the cytosol of all known organisms. It is encoded by gënes belonging to the gënes gapC superfamily. This enzyme is highly conserved during revolution and strictly NAD+-dependent. It is involved in the glycolysis and neoglucogenesis pathways in most bacteria and eukaryotes (Harris and Waters, 1976). It enables the conversion of one molecule of G3P into 1,3-DPG and the reduction of NAD' into NADH, according to the reaction diagram below.

$$G3P + Pi + NAD^+ \rightleftharpoons 1,3\text{-}DPG + NADH + H^+$$

IV. 1. 2. NADP+-dependent phosphorylating GAPDH (EC 1.2.1.13)

The NADP phosphorylating GAPDH' -dependent encoded by genes related to the gapA gene has been isolated and characterised in several plants (Slaughter and Davies, 1968; Preiss and Kosuge, 1970; Yonushot et *al.*, 1970), algae (Pupillo, 1972; Vacchi et *al.*, 1973; Grisson and Kahn, 1975), cyanobacteria (Valverde et *al.*, 1997) and bacteria (Fillinger et *al.*, 2000).

In unicellular algae, the quaternary structure of this enzyme is a homotetramer (A4) (Li et *al.*, 1997). In higher plants the ancestral gene is duplicated in two genes gapA and gapB (Meyer-Gauen et *al.*, 1994 and 1998) giving rise in the chloroplast to a heterotetramer with structure A2B2 and a homotetramer (A4). The heterotetramer is the majority form. The gapB product (43 kDa) differs from gapA (37 kDa) by incorporating a sequence of 30 amino acids at the carboxyl end (C-terminal), which is responsible for heterotetramer formation and has a regulatory function (Li and Anderson, 1997; Fermani et *al.*, 2007).

In bacteria, the gapA gene is frequently grouped with other glycolytic genes (Schlaepfer et Zuber, 1992; Branny et *al.*, 1998; Eikmanns, 1992). It is also found as an individual gene in *E. coli* (Charpentier et *al.*, 1998), *Synechocystis* (Kaneko et *al.*, 1996) and *Streptomyces aureofaciens* (Komanec et *al.*, 1997).

The different NADP' -dependent GAPDHs have strong sequence similarities with the NAD' - dependent GAPDHs. All the essential amino acids are conserved.

However, these NADP'-dependent enzymes show greater similarities with the phosphorylating NAD+-dependent GAPDHs of bacteria than with the cytosolic GAPDHs of eukaryotes (Martin and Cerff, 1986).

The NADP phosphorylating GAPDH' -dependent has also been found in cyanobacteria where it is encoded by the *gap* 2 gene. It is capable of using NAD+ or NADP+ as a cofactor (Hood and Carr, 1969; Tamoi et *al.*, 1996; Figge et *al.*, 1999; Delgado et *al.*, 2001). The presence of a phosphorylating GAPDH capable of using both NAD+ and NADP' has been demonstrated in the bacterium *Bacillus subtilis* (Kunst et *al.*, 1997; Fillinger et *al.*, 2000).

This protein is present in the stroma of the chloroplasts of all photosynthetic organisms, mainly involved in the RPP pathway (pentose phosphate reductive cycle). It is involved in the process of assimilating CO_2 and converting it to carbohydrate during the Calvin cycle (Cerff, 1982; Brinkmann et *al.*, 1989). It is NADP' -dependent and catalyses a reaction similar to that carried out by NAD' -dependent GAPDH, with the exception of using NADP' instead of NAD' as a cofactor:

$$G3P + P_i + NADP^+ \rightleftharpoons 1,3\text{-}DPG + NADPH + H^+$$

IV.1.3 Non-phosphorylating NADP+-dependent GAPDH (EC 1.2.1.9)

The NADP non-phosphorylating glyceraldehyde-3-phosphate dehydrogenase' -dependent (EC 1.2.1.9, GAPDHN) was first detected in the cytosol of algae and plants. It has also been described in bacteria such as *Streptococcus mutans* and *Streptococcus salivarius* (Boyd et *al.*, 1995; Habenicht, 1997).

It is also called G3P: NADP' oxido-reductase and directly converts G3P to 3-PG, without incorporating inorganic phosphate, in the presence of the NADP cofactor' and a molecule of water (Iglesias et Losada, 1988). It also prevents phosphorylation of the substrate and consequently the production of one molecule of ATP by the 3-PG kinase (Iglesias et *al.*, 1987). This reaction is irreversible and causes ionisation of the medium, releasing two protons (H'):

$$G3P + NADP^+ + H_2O \longrightarrow 3\text{-}PG + NADPH + 2H^+$$

IV.2 Physiological role of GAPDH

GAPDH catalyses the reversible oxidative phosphorylation of G3P to 1,3-DPG, in the presence of inorganic phosphate and using nicotinamide adenine nucleotide (NAD') as an electron acceptor. It plays an essential role in the metabolic pathways of glycolysis or neoglucogenesis. This glycolytic enzyme, once consideredëröe as simply a classic mëtabolic protëin involved in energy production, is instead a multifunctional protein with defined functions in many intracellular processes in addition to its glycolytic function (Sirover, 2005).

Depending on its cellular location (cytosolic, membrane or nucleus), GAPDH is thought to be involved in the mechanisms of endocytosis and membrane fusion (Glaser and Gross, 1995), in transport by vesicular secretion (Tisdale, 2001), in the control of translation, in the mechanisms of nuclear transport of transfer RNA, in DNA replication and repair (Meyer-Siegler et *al.*, 1991) and in cell death (Hara et *al.*, 2005).

In the cytoplasm, GAPDH exists mainly as a tetramer composed of 4 identical subunits, each with a catalytic thiol group. While this cytoplasmic enzyme continues to retain its fundamental role as a glycolytic enzyme, several studies have shown that certain post-translational modifications of this enzyme lead it to functional pathways other than glycolysis (Figure 10). The oxidative inhibition of GAPDH by S-thiolation (a reversible reaction) is a controlled response that allows cells to redirect their carbohydrate flow from glycolysis to the pentose phosphate pathway, generating NADPH whose reducing power inside the cells protects them from oxidative stress (Ralser et *al.*, 2007).

Other studies have shown that due to the redox sensitivity of its catalytic cysteine residue, GAPDH can modulate cell signalling pathways in response to oxidative stress (Morigasaki et *al.*, 2008). For example, GAPDH has been shown to bind physiologically to inositol 1,4,5-triphosphate receptors, releasing NADH in the vicinity of the Ca^{2+} channel, thereby regulating intracellular signalling (Patterson et *al.*, 2005) (Figure 10).

On the other hand, nitric oxide-induced stress leads to reversible S-nitrosylation of the cysteine residue (-SNO) of GAPDH (Foster et *al.*, 2009), which facilitates binding of GAPDH to Siah (E3 ubiquitin ligase). Translocation of the GAPDH-Siah complex to the nucleus then occurs, leading to irreversible sulphonation (-SO3H) of GAPDH. This modification can stimulate a "gain of function" that could lead to cell dysfunction or apoptosis (Hara et *al.*, 2005). Post-translational modifications (from S-nitrosylation to sulphonation) involve GAPDH in an irreversible signalling cascade that begins in the cytosol and migrates to other cellular compartments. The regulatory mechanisms of this cascade are important for cellular homeostasis. A protein called GOSPEL has been shown to play a key regulatory role for GAPDH. Under conditions of nitrosative stress, the GOSPEL protein is rapidly S-nitrosylated and retains GAPDH in the cytoplasm. The GAPDH-GOSPEL complex competitively prevents the cytotoxic interaction of GAPDH with Siah (Sen et *al.*, 2009) (Figure 10).

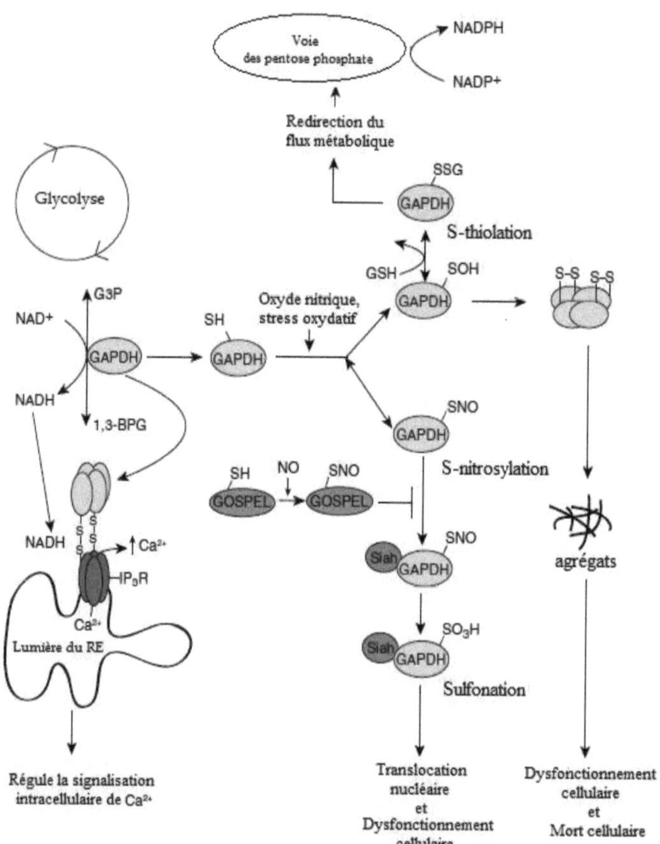

Figure 10. Physiological roles of GAPDH
(after Tristan et al., 2011)

CHAPTER 2

MATERIALS AND METHODS

I. MATERIAL

I. 1 Biological material and growing conditions

The protozoan *Tetrahymena thermophila* (strain SB1969) used as a cellular model throughout this study was kindly provided by the laboratory of Professor Juan Carlos Guttierez at the *Universidad cumpletense de Madrid*. This microorganism is stëriled at 32°C without stirring for 72 h in liquid medium (PPYE) containing 1.5% (w/v) prerteose peptone and 0.25% (w/v) yeast extract (Pousada et *al.*, 1979). The cultures were inoculated with 1% (v/v) from a preculture prepared under the same conditions.

To avoid bacterial contamination, a systematic control is applied to the precultures prepared before use. This control includes a naked eye clouding test, a characteristic smell test for the presence of bacteria and a microscopic observation test to ensure that the precultures are not contaminated with *T. thermophila*.

I. 2. Reagents

I. 2. 1 Hydrogen peroxide and sodium nitroprusside

The hydrogen peroxide (H_2O_2) and sodium nitroprusside (SNP) (Figure 11) used in our work to create oxidative stress and nitrosative stress were sourced from Fluka and Sigma-Aldrich respectively. All solutions of these two stress agents were freshly prepared before each use.

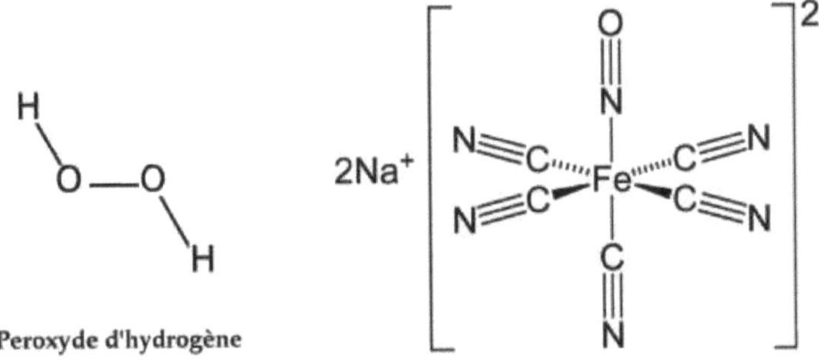

Figure 11. Chemical structures of hydrogen peroxide (H_2O_2) and sodium nitroprusside (SNP)

Table 1. Characteristics of the equipment used

Devices	Features
Autoclave	Brand SANOclav, model Robert-Bosch-Straee 13
Electrophoresis tank	Brand Bio-Rad, model Mini-Protean 3 cell
Electrotransfer tank	Brand Bio-Rad, model Mini Trans-Blot cell
Generator	Brand Bio-Rad, model PowerPac Basic Supply
Scales	- Denver Instrument scale (S-603), sensitë ± 0,0001 g
	- KERN balance (440-33N), sensitivity ± 0.01 g
	- Sigma refrigerated centrifuge, model 2-16K
Centrifuges	- Eppendorf refrigerated centrifuge, model 5415 R - Eppendorf

	MiniSpin Microcentrifuge
Ovens	- BINDER Incubator
	- Memmert drying oven, model U10
Sonicator	Brand Bandelin sonoplus (HD 2070)
Laminar flow hood	Brand Telstar, model V-30/70
Microscopes	- Optical microscope A.KRUSS OPTRONIC
	- Microscope paired with a FinePixe A400 digital camera
pH meter	Brand HANNA instruments, pH 209
Spectrophotometers	- Jenway UV/Visible spectrophotometer (6405)
	- Schimadzu UV/Visible spectrophotometer (UV-1700)
	- Jenway Visible Spectrophotometer (7315)

I. 2. 2 D-gfyceraldehyde-3-phosphate

D-glycëraldehyde-3-phosphate, ий^ё as a substrate for GAPDH, is prepared from the barium salt of D-glycëraldehyde-3-phosphate dïëthylacëtal. The latter is obtained from Sigma-Aldrich.

I. 2. 3 Bradford reagent

Bradford's reagent, used for the dëtermination of the quantity of protëines in ëchantiïIons analyses (Bradford, 1976), is obtained from Merck-Millipore.

I. 2. 4 Protein markers

The molecular weight markers (*Precision Plus Protein Standards*), like the *pI* markers (*IEF Standards*), are obtained from Bio-Rad.

I. 2. 5 Other reagents

The remaining chemical reagents were all obtained from Sigma-Aldrich.

I. 3 Equipment

The makes and modëles of the various devices used in our work are shown in Table 1.

II. *IN VIVO* STUDY METHODS

II. 1 Induction of oxidative and nitrosative stress

Two stress agents were used to induce oxidative and nitrosative stress in the protozoan *T. thermophila*.

II. 1. 1 Induction of oxidative stress by hydrogen peroxide

In order to induce oxidative stress, *T. thermophila* was incubated in the presence of hydrogen peroxide ($H2O2$). H2O2 was added to the protozoan culture medium (PPYE) prior to inoculation, at concentrations determined for each experiment.

II.1.2 Induction of nitrosative stress by sodium nitroprusside

Sodium nitroprusside (SNP) was used to create nitrosative stress. This nitric oxide (NO) donor is used in both *in vivo* and *in vitro* studies. In *in vivo studies*, SNP is added to the PPYE culture medium at the start of the experiment prior to inoculation, at the desired concentrations.

II.2 Assessment of the effect of exposure to oxidative and nitrosative stress

II.2.1 Effect on the growth of T. thermophila

The effect of stress agents on cell growth was ëvaluatedë by incubating *T. thermophila* with different concentrations of H2O2 or SNP. These reagents are addedës to the culture media from freshly prepared mëre solutions (97.8 mM of H2O2 and 0.5 M of SNP). H2O2 was added from 0.05 to 1.5 mM in 50 ïM steps, while SNP was added from 0.05 to 15 mM in 100 ЦM steps. 500 ml Erlenmeyer flasks, containing 100 ml of PPYE medium and the stress agents (H2O2 or SNP) at different concentrations, were aseptically inoculated with 1% (1.5 x 10^5 cells/ml) of a *T. thermophila* preculture. After 72 h incubation at 32°C without agitation, cell counts were carried out under an optical microscope using a haemocytometer (Malassez cell). Simultaneously, a control was carried out under the same conditions in the absence of stress agents.

Determination of the minimum inhibitory concentrations (MICs) of H2O2 and SNP was based on the turbidity observed in the culture medium after 72 h incubation of the protozoa at 32°C in the presence of different concentrations of the stress agents. The lowest concentration showing no turbidity in the medium to the naked eye was taken as the MIC. The concentrations of H2O2 and SNP inhibiting the growth of *T. thermophila* by 10% (IC10) and 50% (IC50) were estimated by probit analysis (Bliss, 1935).

The effect of H2O2 and SNP on the different phases of *T. thermophila* growth was evaluated in the presence of each of the stress agents at the 50% inhibitory concentration (IC50) for 140 h at 32°C. A control was carried out in the absence of the stress agent. At different time intervals, 1 ml aliquots were taken from each culture and fixed with neutral formalin buffer (10% (v/v) formalin in phosphate-buffered saline pH 7 (PBS)) for one hour. These aliquots were counted according to the protocol described above. All experiments were repeated at least three times.

II. 2. 2 Effect on the number of generations and generation time

To assess the influence of the two stress agents on the number and time of generation, cultures of *T. thermophila* in the absence and presence of H2O2 or SNP at the inhibitory concentrations IC 10 and IC50 were maintained for 72 h at 32°C. Aliquots of 1 ml were taken from each culture, diluted in distilled water and fixed as described previously (II.2.1). Cells were counted under a microscope. The generation number and generation time were calculated according to the following formulae (Dias et *al.*, 2003):

$$Nombre\ de\ génération\ (n) = \frac{(\log N_1 - \log N_0)}{\log 2}$$

$$Temps\ de\ génération\ (g) = \frac{temps\ de\ croissance}{nombre\ de\ génération}$$

$$Generation\ time\ (g) = \frac{growth\ time}{number\ of\ generations}$$

N1 being the number of cells at 72 h and N0 the number of cells at T0. growth time = 72 h.

II. 2. 3 Effect on morphology

In order to assess the effects of stress agents (H2O2 and SNP) on *T. thermophila* cell morphology, these agents were addedë séparëment to the protozoan culture medium at IC50 inhibitory concentrations. After 72 h of incubation at 32°C, the morphological changes affecting the treated cells were observed using an optical microscope coupled to a digital camera allowing photographs to be taken with amplitudes of *20 and *40.

II. 2. 4 Determination of enzyme activities

One of the parameters used to assess the response to oxidative and nitrosative stress induced by stress agents is the measurement of the activities of enzymes directly or indirectly involved in the cellular defence system.

II. 2. 4. a- Preparation of the crude protein extract

Cultured cells were recovered by centrifugation at *6000g* for 15 minutes at 4°C. After removal of the supernatant, the pellet containing the recovered cells was washed 3 times with a solution of Tris HCl (20 mM, pH 7.5).

The crude protein extract is obtained after breaking down the cell membrane using ultrasound. This mechanical lysis process breaks up the cells and releases their cytoplasmic contents. The cells are suspended in a lysis buffer (50 mM Tris-HCl pH 7.5 containing 1 mM ethylene-diamine-tetra-acetic acid (EDTA), 10 mM 2-e-mercaptoethanol, 1% (v/v) glycerol and 1 mM phenylmethanesulphonyl fluoride (PMSF)) at a rate of 3 ml/g. The presence of EDTA and PMSF serves to inhibit the action of proteases, and 2-e-mercaptoethanol is required to maintain a reducing environment. The cell suspension is then ultrasonicated for 4 min at 90% power (12 cycles of 20 s each, alternated with 1 min rest in an ice bath to avoid heating the sample, which can damage the integrity of the extracted cell proteins). The solution is then clarified by centrifugation at 15,000 g for 45 min at 4°C. The supernatant obtained is considered to be the crude protein extract.

II. 2. 4. b- Determination of catalase activity

Catalase activity is determined using the technique described by Aebi (1984). This technique is based on monitoring the decomposition of H_2O_2 into H_2O and O_2 following the catalytic activity of catalase.

The crude extract is added to a quartz cell containing 50 mM phosphate buffer (K_2HPO_4/KH_2PO_4, pH 7). A pre-incubation period of 2 min was carried out at 30°C. The cuvette was then placed in a spectrophotometer. The reaction was initiated by the addition of H_2O_2. The decomposition of 7.5 mM H_2O_2 was determined directly by decreasing the absorbance at 240 nm. A unit of catalase activity is defined as the amount of enzyme required to decompose 1 |imol of H_2O_2 in 1 min at pH 7.0 and 25°C. The activity of the enzyme is expressed as an activity unit/mg of protein.

II. 2. 4. c- Determination of superoxide dismutase activity

The technique for determining SOD activity is that described by Paoletti et *al* (1986). This technique is based on the inhibition of NADH by superoxide dismutase. The decrease in the rate of NADH oxidation is proportional to the concentration of the enzyme.

In the absence of SOD, the auto-oxidation of 2-e-mercaptoethanol in the presence of EDTA/$MnCl_2$ will generate superoxide anions (O_2^-) in the reaction medium and cause the oxidation of NADH to NAD^+, resulting in a decrease in absorbance at 340 nm.

In the presence of SOD, there is competition between the generation (2-e-mercaptoethanol + EDTA/$MnCl_2$) and dismutation of superoxide anions. This tends to reduce the quantity of superoxide anions in the medium, inhibiting NADH oxidation and therefore reducing absorbance. It is estimated that 50% inhibition corresponds to one unit of enzyme.

The reaction is based on mixing 5 mM EDTA, 2.5 mM $MnCl_2$, 0.27 mM NADH, 3.9 mM 2-e-mercaptoethanol in 50 mM phosphate buffer (pH 7.0) with 50 Щ of the crude extract. The reaction is initiated by addition of NADH to the final concentration of 0.27 mM. Measurement of activity is determined by spectrophotometry at 340 nm.

II. 2. 4. d- Determination of lipid peroxidation

The rate of lipid peroxidation was quantified in terms of thiobarbituric acid reactive substances (TBARS) using the method of Samokyszyn and Marnett (1990). One millilitre of the crude extract is added to 1 ml of a solution consisting of 0.375% thiobarbituric acid and 15% trichloroacetic acid in 0.25 M hydrochloric acid. The mixture was then placed in a water bath at 100°C for 15 min, then cooled in ice to stop the reaction. Centrifugation was performed at *1000g* for 10 min and the optical density of the supernatant was measured at 535 nm. The dëgradation product of polyunsaturated fatty acids was calculated using the extinction coefficient of 1.56×10^5 M^{-1} cm^{-1} .

II. 2. 4. e- Determination of GAPDH activity

The enzymatic activity of the NAD phosphorylating GAPDH$^+$ -dependent is determined by spectrophotometry at 30°C by measuring the appearance of NADH at 340 nm (Iglesias et *al.*, 1987; Serrano et *al.*, 1993). The enzyme preparation is added to a reaction mixture containing tricine-NaOH buffer (50 mM; pH 8.0), sodium arsenate (Na2HAsO4) (10 mM), NAD$^+$ (1 mM) and D-G3P (2 mM). D-G3P was obtained by acid hydrolysis of the salt of D-glyceraldehyde-3-phosphate diethylacetal according to the supplier's (Sigma-Aldrich) instructions. The total volume of the reaction mixture is 1 ml. A unit of enzyme activity is defined as the amount of enzyme catalysing the reduction of 1 |imol of NAD$^+$ per minute. All experiments and measurements were performed three times.

II. 2. 4. f- Protein determination using the Bradford method

The Bradford method is a rapid and highly sensitive method for measuring protein concentration (Bradford, 1976). This method is based on a colorimetric reaction between proteins and a dye: coomassie blue G-250. This reagent, which is brick-red in the free state, takes on a blue hue when bound to proteins. To apply this assay technique to a protein solution of unknown concentration, a "DO = f (protein concentrations)" calibration line must first be drawn under identical experimental conditions, using standard protein solutions of known concentrations.

BSA (bovine serum albumin) is used as the standard protein in a concentration range from 0 to 20 |ig. A sample volume of 5 to 10 |il is reduced to 800 |il with 1 bi-distilled water. Two hundred microlitres of Bradford reagent (25 mg coomassie blue G-250, 12.5 ml absolute ethanol, 25 ml phosphoric acid 85%, qsp 250 ml distilled water) are added. The tubes develop a coloration ranging from brick red at low protein concentrations to blue at high concentrations. Depending on the protein concentration of the samples, prior dilutions may be necessary to ensure that the absorbance values of the samples are within the range of the absorbance values of the standard range. The optical density of the developed colour is then measured at 595 nm.

Knowing this absorbance, the unknown protein concentration can be determined by reference to the calibration curve plotted. The protein concentration of each sample is deduced taking into account the initial dilution factor.

II.3 Evaluation of the antioxidant potential of certain essential oils

II.3.1 The essential oils used

Essential oils extracted from 7 aromatic plants were used to assess their anti-stress effects. These essential oils were obtained from Professor Emna Ammar (Ecole Nationale des Ingënieurs de Sfax, Tunisia). They have been kept at 4°C and protected from light until use. These essential oils were extracted, by hydrodistillation, from lavender (*Lavandula angustifolia*), geranium (*Pelargonium robertianum*), thyme (*Thymus capitalus*), rosemary (*Rosmarinus officinalis*), cypress (*Cupressus sempervireus*), juniper (*Juniperus phoenica*) and clove (*Syzygium aromaticum*) (Ben Saida, 2007).

II. 3.2 Determination of the MIC of essential oils

The minimum inhibitory concentration (MIC) is used to assess the sensitivity of the protozoa to essential oils. This concentration is determined using the two-series cascade dilution method. Each essential oil was dissolved in dimethylsulphoxide (DMSO). Series of dilutions ranging from 10^{-1} to 10^{-6} were then prepared. Five microlitres of each dilution was added to 5 ml of PPYE medium in test tubes. An inoculum of *T. thermophila* (1.5×10^5 cells/ml) was added to each tube containing the culture medium and essential oil. A tube without essential oil containing DMSO (0.1% v/v, amount with no effect on *T. thermophila*) was used as a control. Growth of the protozoa was observed with the naked eye after 72 h of exposure to essential oils at 32°C. The MIC (the lowest concentration for which no growth is visible with the naked eye) was then determined.

II.3.3 Determination of the antioxidant potential of essential oils

To assess the effect of essential oils on oxidative and nitrosative stress, *T. thermophila* was incubated in the presence of IC50 stress agents (0.7 mM H2O2 or 1.8 mM SNP) in PPYE medium supplemented with essential oil at the non-toxic dilution of 10^{-9} . The stress agents and essential oils were added to the culture medium simultaneously prior to inoculation with *T. thermophila*. Growth was monitored, as described above, by counting cells at different time intervals. Controls were carried out in the presence of the stress agent and in the absence of the essential oil.

II.3.4 Synergistic effect of essential oils

The essential oils selected to assess the synergistic effect were those extracted from lavender, thyme and geranium. A mixture of two essential oils was then added, at the non-toxic dilution of 10^{-9} , to the *T. thermophila* (PPYE) culture medium containing the stress agent (H2O2 or SNP) at IC50. Growth curves were followed, as previously described, by counting cells at different time intervals. Controls were performed on PPYE medium supplemented with the stress agent and inoculated with *T. thermophila*.

III. *IN VITRO* STUDY METHODS

In order to study the *in vitro* effect of stress agents (H2O2 and SNP) on the key carbohydrate metabolism enzyme glyceraldehyde-3-phosphate dehydrogenase (GAPDH), chosen as the modèle, it was first necessary to purify and characterise it.

III.1 GAPDH purification procedure

GAPDH was purified from *T. thermophila* grown for 72 h at 32°C. Purification was carried out by fractional precipitation with ammonium sulphate, followed by two column chromatographies (cation exchange chromatography and anion exchange chromatography). All purification steps were carried out at 4°C.

III.1.1 Fractional precipitation with ammonium sulphate

This technique is based on the differential solubility of proteins. Proteins are separated according to their tendency to precipitate more or less quickly by changing the ionic strength of the solution containing them. The electrolyte most commonly used to precipitate proteins is ammonium sulphate ($(NH4)2SO4$)). This salt is highly soluble in aqueous solution and allows very high ionic strengths to be achieved. Its solubilisation does not affect the temperature of the solution and protects the proteins against denaturation and bacterial growth.

Thus, the crude extract, obtained as described above (II.2.4.a), is added to ammonium sulphate at 55% saturation and stirred gently for 1 h in an ice bath. To avoid surface denaturation, the solution should not be shaken vigorously and the ammonium sulphate should be added gradually. The precipitated proteins were removed by centrifugation at 15,000 g for 45 min at 4°C. The enzyme in the supernatant was then precipitated by adding sufficient ammonium sulphate to reach 88% saturation. The solution was gently stirred overnight at 4°C. The protein fraction precipitated between 55 and 88% saturation was recovered by centrifugation at 15,000g for 45 min at 4°C, then resuspended in a minimal volume of buffer A: 50 mM Tris-HCl pH 7.5, 1 mM EDTA, 10 mM 2-e-mercaptoethanol, 1% (v/v) glycerol and 1mM PMSF.

In order to eliminate the ammonium sulphate, we dialysed the fraction obtained. The enzyme preparation was dialysed 2 times through a dialysis membrane (MWCO 1000) against 5 l of buffer A overnight at 4°C with gentle agitation.

III. 1. 2. Anion exchange chromatography

In an anion exchange column, proteins stick by ëlectrostatic affinity to positively charged groups in the resin (diagram below). Negative ions or acid groups on a protein can interact with such a resin.

(diethyl aminoethyl-)

We used DEAE-cellulose (Fluka, Buchs, Switzerland) as an anion-exchange matrix. The gel was degassed for 30 min before being introduced into a column with an internal diameter of 1.5 cm and a length of 12 cm. The column is then equilibrated by several washes with buffer A. The dialysed iraction was introduced into the column and the enzyme was then eluted using buffer A at a flow rate of 12 ml/h. 1 ml iractions were collected and those with GAPDH activity were pooled.

The column can be reused after regeneration by washing 5 volumes of the column with a solution of NaCl (2 M) and urea (1 M) in Tris-HCl buffer (50 mM, pH 7.5). The column is then washed and re-equilibrated with 50 mM Tris-HCl buffer pH 7.5.

III.1.3 Cation exchange chromatography

Cation exchange chromatography is based on the use of a resin carrying negative groups. In our case, we used a Mono-S resin capable of interacting with cations and basic groups (see diagram below).

The active fractions obtained after chromatography on DEAE-cellulose are deposited on a Mono-S column (HR 5/5) previously equilibrated with buffer A. Unbound proteins are eluted during washing, while bound proteins (GAPDH) are eluted by a linear gradient (0 to 300 mM; total volume 130 ml) of potassium chloride (KCl) at a flow rate of 12 ml/h. Fractions of 1 ml are collected and those containing GAPDH activity are pooled and analysed.

III. 2. Determination of the molecular weight of purified GAPDH

In order to check the homogeneity and purity of the enzyme after the various purification steps and also to determine the molecular weight of the purified enzyme, the polyacrylamide gel electrophoresis technique was used.

III.2.1 Polyacrylamide gel electrophoresis under denaturing conditions

Proteins from the various purification steps are separated by vertical electrophoresis on polyacrylamide gel (mini-gel, 8 x 10 cm) using the method described by Laemmli (1970). The use of a denaturing agent such as sodium dodecyl sulphate (SDS), which comprises a hydrophilic, negatively-charged polar sulphate head and a 12-carbon apolar chain, causes the quaternary structure of the proteins to dissociate into monomeric forms, to which it binds. Under these conditions, the separation of proteins on the gel is based solely on their size, since their charge density is made negative overall by the presence of SDS.

III. 2. 1. a- Preparation of gels

The polyacrylamide concentration of the gel is chosen according to the size of the molecules to be analysed. In our case, we used a 4% polyacrylamide gel as the concentration gel and a 12% polyacrylamide gel as the separation gel. The composition of these two gels is shown in Table 2.

Table 2: Composition of polyacrylamide gels for electrophoresis under denaturing conditions (PAGE-SDS)

Stacking gel (4%)	**Separation gel (12%)** *"Running gel*
- 0.7 ml acrylamide/bisacrylamide (29.2/0.8%)	- 4 ml acrylamide/bisacrylamide (29.2/0.8%)
- 1.25 ml Tris-HCl (0.5 M, pH 6.8)	- 2.5 ml Tris-HCl (1.5 M, pH 8.8)
- 3 ml Distilled water	- 3.35 ml Distilled water
- 50 gl SDS (10%)	- 100 gl SDS (10%)
- 25 gl APS (10%)	- 50 gl APS (10%)
- 5 gl TEMED	- 10 gl TEMED

Gel polymerisation is initiated by the addition of 0.1% (w/v) ammonium persulphate (APS) and N,N,N',N'-tetra-methylene-diamine (TEMED) catalyst (6 mM).

III. 2. 1. b- Sample preparation and migration

A volume containing 50 jug of protein is тёlапдё with 5 Щ of loading buffer whose composition is as follows: 60 mM Tris-HCl (pH 6.8), 1% (w/v) SDS, 10% (v/v) дlусёrоl, 0.01% (w/v) bromophёnol blue and 1% (v/v) 2-B-mercaptoethanol. The mёlange is heated for 5 minutes at 100°C and deposited in the gel wells.

III. 2. 1. c- Migration and revelation

Migration was carried out at room temperature with a constant voltage of 100 V. The migration buffer used was Tris-HCl (25 mM, pH 8.5), glycine (320 mM) and SDS (0.1%, w/v). The gels were stained overnight at room temperature under agitation with a distilled water / methanol / acetic acid solution (5/4/1, v/v/v) containing 0.25% (w/v) coomassie blue R-250. The gel is then washed with the same solution without the coomassie blue. This washing step allows the gel to be decolourised without removing the coomassie blue bound to the proteins.

The molecular weight of the GAPDH subunit was determined by reference to the line (relative mobility as a function of the logarithm of the molecular weight) drawn on the basis of the relative migration of the molecular weight markers (distance of the marker migration relative to the distance of the migration front).

III.2.2 Determination of the molecular weight of GAPDH by FPLC

The molecular weight of native GAPDH was determined by gel filtration through a Superdex H/R column attached to a fast protein liquid chromatography (FPLC) system (Pharmacia, Uppsala, Sweden). Tris-HCl buffer (50 mM, pH 7.5) containing 150 mM NaCl and 5 mM MgCl2 was used as the mobile phase at a flow rate of 0.3 ml/ min. The molecular weight markers used to calibrate the column and construct the calibration curve were ferritin (440 kDa), catalase (232 kDa), aldolase (158 kDa), BSA (67 kDa) and ovalbumin (43 kDa). These proteins were detected in the eluates by absorbance at 280 nm.

111.3. Determination of the isoelectric point of GAPDH

The isoelectric point of purified GAPDH was determined using the isoelectrofocusing (IEF) technique on a polyacrylamide gel according to the procedure described by Robertson et *al.* (1987). The principle of this technique is based on the pH gradient created by the ampholytes introduced into the gel preparation, under high voltage.

111.3.1 Preparation of the 5% polyacrylamide gel

The 5% polyacrylamide gel used to determine the isoelectric point (*pI*) is composed of :

- 2 ml of acrylamide / bisacrylamide solution (29.2 / 0.8%),
- 0.6 ml ampholyte solution (pH range 3.5-10; Pharmalyte 3.5-10),
- 2.4 ml of 50% (v/v) glycërol in distilled water,
- 7 ml distilled water.

The mëlange is dëgazë, and polymërisation of the gel is initiated at room tempërature by the addition of 50 Щ of 10% (w/v) ammonium persulphate and 20 Щ of TEMED. The mëlange is then poured between two glass plates (8 x 10 cm mini-plate) spaced 1.5 mm apart.

III. 3. 2 Preparation of samples

The equivalent of 40 jugs of protëines is mixed with an equal volume of a solution composed of glycërol (80%, v/v) and ampholytes (4%, v/v) in the same pH range used for gel preparation (3.5-10).

III. 3. 3 Migration and revelation

Solutions of 25 mM sodium hydroxide (NaOH) and 20 mM acëtic acid are used as electrode solutions at the cathode and anode respectively. Migration of the protëines is carried out at 4°C initially for 2 h at 200 V, and then for 2 h increasing the voltage to 400 V. After migration, the gel is washed once with a 10% (w/v) TCA solution and then several times with the same 1% (w/v) solution before being thoroughly rinsed with distilled water. Proteins are visualised on the gel after staining with coomassie blue as described previously (III.2.1.c).

The proteins used as *pI* markers are in the pH range 4.4-9.6. The *pI of* purified GAPDH is determined from the calibration curve representing the distance travelled by each marker protein as a function of its *pI*.

III. 4 Determination of the cinetic parameters of purified GAPDH

The method used to determine the enzyme's cinetic constants is that described by Florini and Vestling (1957), which involves varying the concentration of one substrate, NAD^+ (from 0.02 to 0.32 mM) or D-G3P (from 0.02 to 0.064 mM), and fixing that of the other substrate (NAD^+ or D-G3P). Activity measurements were carried out under the same conditions as described above (II.2.4.e), using tricin buffer (50 mM, pH 8.0) at 30°C. The maximum speed of the enzymatic reaction (v_{max}) and the Michaelis-Menten constants of each substrate (K_m) were determined graphically using the Lineweaver-Burk (1934) representation.

III. 5 Determination of physico-chemical parameters

III.5.1 Influence of temperature on GAPDH activity

The effect of temperature on the activity of purified GAPDH was determined by monitoring the activation and denaturation of the enzyme.

III. 5. 1. a- Thermal activation of GAPDH

The activation tempërature a ële dëterminëe by measurement of GAPDH activity in tricine buffer (50 mM, pH 8) heated to the dësirëe tempërature in a range from 5 to 65°C. The reaction mixture is the same as described previously (II.2.4.e). The reaction is dëclenchëe by the addition of the purified enzyme.

III. 5. 1. b- Thermal denaturation of GAPDH

Thermal dënaturation of GAPDH is ëtudiëed by pre-incubating the enzyme for 10 min at the tempërature dësirëe in the range of 5 to 65°C. The reaction is started by adding the preheated enzyme to the usual reaction mixture.

III.5.2 Influence of pH on GAPDH activity

The influence of pH on GAPDH activity was ëtudiëed in the pH range 4 to 9.5 using a mixture of buffers with different pKa (50 mM Tris-HCl, 50 mM MES, 50 mM HEPES, 180 mM sodium acetate and 50 mM phosphate). The buffer was aliquoted and pH adjusted to different values (from 4 to 9.5) using 1 M HCl solution or 1 M NaOH solution. GAPDH activity was measured at the different pH values as described previously (II.2.4.e).

III.6 Production of polyclonal anti-GAPDH antibodies

To produce polyclonal antibodies to GAPDH purified from *T. thermophila*, a 1.5 kg albino rabbit was injected subcutaneously at different sites with 500 |ig of the enzyme mixed with Freund's incomplete adjuvant (v/v). Prior to injection, 10 ml of the rabbit's blood was collected to recover the pre-immune serum. After three weeks, a second booster injection was given, and one week later another booster was given. One week after this last booster, a 60 ml volume of blood was drawn from the immunised rabbit. The antiserum was separated by allowing the collected blood to coagulate for 1 hour at 30°C and then overnight at 4°C. After centrifugation at 4000 g for 15 min, the serum obtained containing polyclonal anti-GAPDH antibodies was supplemented with sodium azide (0.02%) and stored at -20°C until use.

III.7 Western blot analysis

III.7.1 Electrotransfer on nitrocellulose membrane

Proteins are transferred using a semi-dry transfer device. After polyacrylamide gel electrophoresis in the presence of SDS, the proteins are transferred to a 0.45 μm nitrocellulose membrane (Schleicher and Schuell, Dassel, Germany) previously equilibrated in transfer buffer: Tris-HCl (25 mM, pH8.3), glycine (192 mM), SDS (1.3 mM) and methanol (20%, v/v).

Transfer is achieved by applying a current of 200 mA for 45 min. The *sandwich* assembly comprises a sheet of nitrocellulose placed under the polyacrylamide gel. The assembly is placed between Whatman paper and brought into contact with the two graphite plates.

The membrane was then stained for 5 min with 0.2% (w/v) ponceau red in a 3% (w/v) TCA solution and washed thoroughly with distilled water. The appearance of red coloured bands on the membrane tëmotifies the presence of the protëines and therefore the efficiency of the transfer.

III.7.2 Immunodetection

The membrane containing the transfërëed protëins is incubated overnight at room temperature with gentle agitation in a solution of 5% (w/v) BSA prepared in TBS buffer (Tris-HCl (20 mM, pH 7.5) containing 150 mM NaCl). This ëtape allows the saturation of non-spëcific sites to subsequently prevent any non-spëcific fixations. The membrane is then incubated, in the presence of immune sërum at a dilution of 1:500 in TBS, for 2 h at room temperature with stirring. The membrane is then washed 3 times (15 min each) with TBS and a fourth 15 min wash with TBST (TBS containing 0.05% (v/v) Tween-20), then incubated for 45 min in the presence of the secondary antibody (rabbit anti-IgG conjugated to peroxidase (Promega, Madison, USA)) at a dilution of 1/1000 in TBST. The membrane is then washed 3 times with TBST (15 min each) and rinsed with TBS for 15 min. Revelation is carried out in the dark in 20 ml of a solution consisting of a mixture of 4-chloro-1-naphthol (12 mg in 4 ml of mëthanol), TBS (16 ml) and H2O2 (25 gl of a 33% (v/v) solution in water). As soon as the dark bands appear on the membrane, the reaction is stopped by washing with running water. The revised membrane is then dried between two Whatman papers.

III.8 Effect of oxidative and nitrosative stress on the activity of purified GAPDH

To assess the *in vitro* effect of oxidative and nitrosative stress on the enzymatic activity of *T. thermophila* GAPDH, the purified enzyme was incubated in the presence of different concentrations of H2O2 and SNP. The H2O2 concentration range used was between 0 and 25 mM, while the SNP concentration range was between 0 and 75 mM. These ranges were prepared in Tris-HCl buffer (25 mM, pH 8.0) containing 1 mM EDTA. The enzyme was incubated for 30 min at 30°C in the presence of each concentration of stressor. GAPDH activity was measured by spectrophotometer at 340 nm.

The GAPDH activity inhibitory concentrations (IC10 and IC50) of each stressor, defined as the concentrations that inhibit enzyme activity by 10% and 50% respectively, were calculated using probit analysis (Bliss, 1935).

IV. STATISTICAL ANALYSIS

The results prësentës in our work are expressedës as the mean of the values obtained from at least three indëdependent expëriences. Mean values are given with their standard ëstandard deviations (SEM). Student's *t-test* was used for comparison of means. The statistical significance thresholds are: $p < 0.05$ (significant) and $p < 0.01$ (highly significant).

CHAPTER 3

RESULTS AND DISCUSSION

PART 1: EFFECT OF OXIDIZING AND NITROSATIVE STRESS ON THE GROWTH, MORPHOLOGY AND PHYSIOLOGY OF *Tetrahymena thermophila*

In this first part, we present the results concerning the effects of oxidative stress and nitrosative stress on the ciliated protozoan; *T. thermophila*. Hydrogen peroxide (H_2O_2) and sodium nitroprusside (SNP) were the two stress agents chosen and used to assess the impact of these two types of stress on the growth, morphology and physiology of *T. thermophila*.

I. EFFECT OF OXIDATIVE AND NITROSATIVE STRESS ON THE GROWTH OF *T. THERMOPHILA*

To assess the effect of oxidative and nitrosative stress on the growth of *T. thermophila*, the protozoan was incubated at 32°C for 72 h at different concentrations of H2O2 (from 0.05 to 1.5 mM) and SNP (from 0.05 to 15 mM). The control culture, carried out under the same conditions but in the absence of stress agents, indicated that *T. thermophila* was in its logarithmic growth phase after 72 h incubation.

The cell count results for the different cultures, shown in Figure 12, indicate that *T. thermophila* growth is sensitive to the stress agents tested. Sensitivity depends on the type of stress and the concentration used. The dose-response curves differ according to the nature of the stressing agent.

For H2O2 at concentrations below 0.5 mM, no significant change in cell number was observed (Figure 12a). Growth inhibition only became significant from 0.5 mM H_2O_2 upwards. In the case of SNP (Figure 12b), there was a negative correlation between cell number and concentration; as SNP concentration increased, cell number decreased. As shown in Figure 12, H2O2 and SNP completely inhibited the growth of *T. thermophila* at 1 and 10 mM respectively.

Probit analysis (Bliss, 1935), used to calculate inhibitory concentrations (IC_{10} and IC_{50}), revealed that H2O2 is more toxic than SNP with lower IC_{10} and IC_{50} values (IC_{10} H2O2 = 0.499 mM < IC_{10} SNP = 0.534 mM) (IC_{50} H2O2 = 0.705 mM < IC_{50} SNP = 1.845 mM) (Table 3).

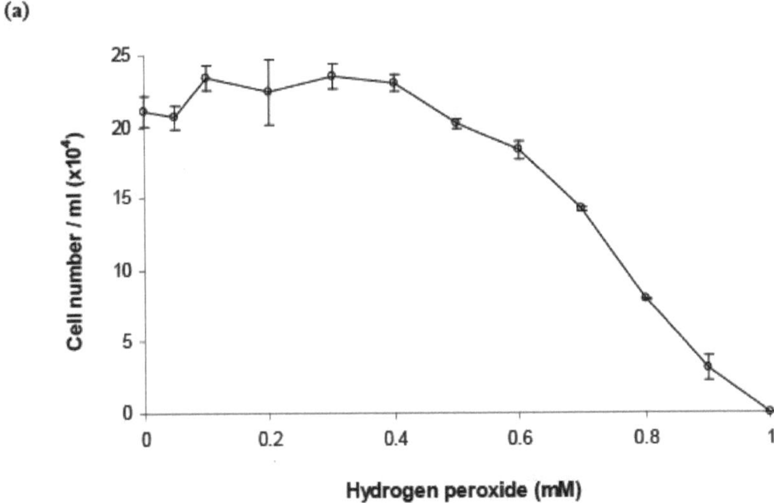

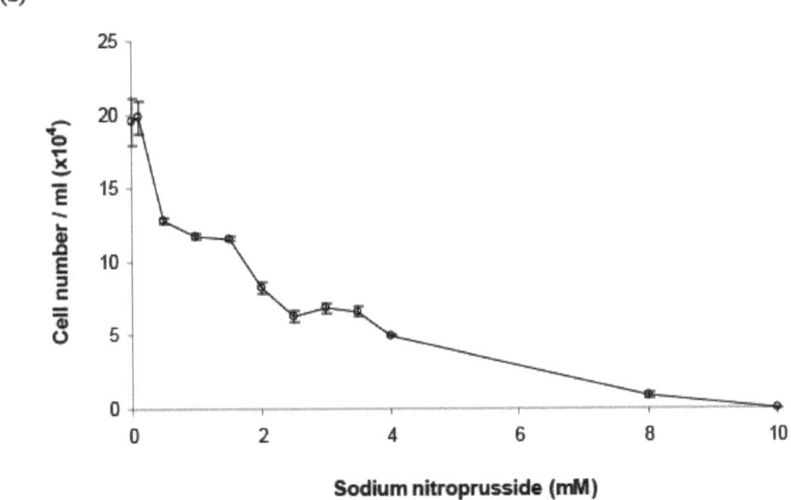

Figure 12. Effects of H2O2 (a) and SNP (b) on the growth of *T. thermophila*

The number of *T. thermophila* cells was determined after 72 h incubation at 32°C in the presence of different concentrations of stress agents. Data represent the means of 3 separate experiments ± standard deviations.

Table 3. Inhibitory concentrations (IC10, IC50 and MIC) of H2O2 and SNP on the growth of
growth of _T. thermophila_

	IC10 (mM)	IC50 (mM)	MIC (mM)
H2O2	0,499 ± 0,014	0,705 ± 0,009	1
SNP	0,534 ± 0,067	1,845 ± 0,084	10

The values represent the averages of 3 separate experiments ± standard deviations.

To determine how stress agents affect the number and generation time of _T. thermophila_, we compared untreated cells with those treated with H2O2 and SNP. The results obtained indicate that the normal number and generation time of _T. thermophila_ after 72 h of growth, under our culture conditions, were 8.18 and 8.79 h respectively. The addition of stress agents influenced both parameters; the generation number decreased and the generation time increased significantly (p<0.05) (Table 4).

Table 4. Number and generation time of _T. thermophila_ incubated at 32°C for 72 h in the presence of H2O2 and SNP at IC10 and IC50.

		Number of generations	Generation time (h)
Timer		8,19 ± 0,09	8,79 ± 0,10
H2O2	_IC10_	8,14 ± 0,02	8,85 ± 0,03
	_IC50_	7,64 ± 0,02*	9,43 ± 0,03*
SNP	_IC10_	7,48 ± 0,04*	9,62 ± 0,05**
	_IC50_	6,81 ± 0,05**	10,57 ± 0,07**

*Each value represents the mean of 3 separate experiments ± standard deviation. *
Significantly different from control at p< 0.05
*** Significantly different from the control value at p< 0.01*

II. EFFECT OF H2O2 AND SNP ON THE MORPHOLOGY OF _T. THERMOPHILA_

In another series of experiments, _T. thermophila_ cells were cultured at 32°C for 72 h in the presence of a stress agent (H2O2 or SNP) at IC50 in order to assess the effects on cell morphology.

The results of this analysis présentés in Figure 13 show that changes in morphology ët were more pronounced in SNP-treated cells than in H2O2-treated cells. The morphology of SNP-treated cells was markedly altered compared with the control. The cells took on various abnormal shapes and their mobility was also affected. They turned in circles or moved irregularly. In addition, we observed cells that had initiated cell division but had not completed cytokinesis (Figure 13c). In the presence of H2O2, _T. thermophila_ cells increase in size by enlarging their internal volume. A clear decrease in cell number was also observed at the IC50 of each stress agent.

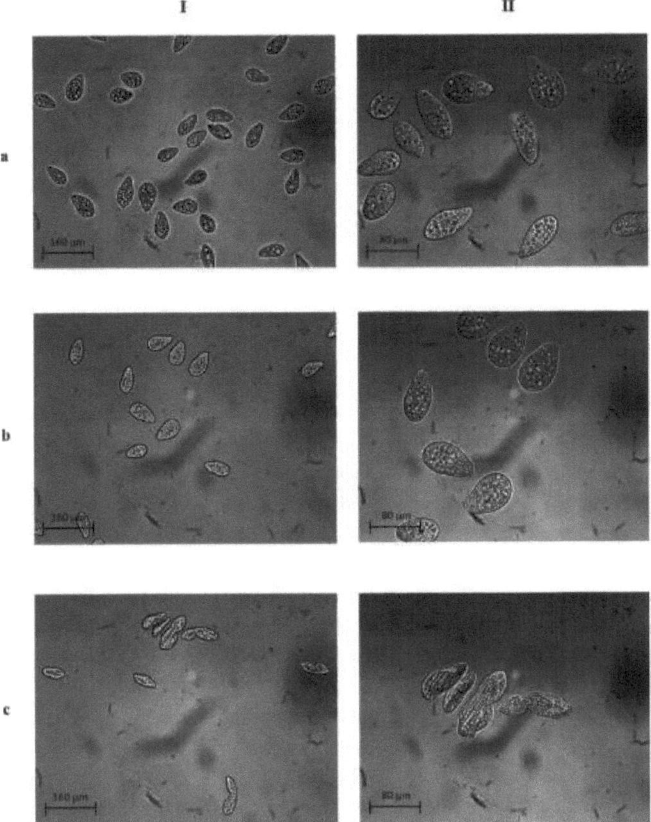

Figure 13. Microscopic analysis of *T. thermophila* cells grown at 32°C for 72 h in the absence of stress agent (a) and in the presence of H2O2 (b) and SNP (c) at IC50

Microscopic images were taken at two magnifications (I: x20 and II: x40).

We ëtudiedë the reversibility of these effects. *T. thermophila* cells were grown for 72 h at IC50 of H2O2 or SNP, then inoculated into freshly prepared stress-free culture medium for 72 h at 32°C. The results showed that the cells were able to recover their normal morphology and growth capacity. Cells treated with H2O2 and SNP regained their normal shape.

III. PHYSIOLOGICAL EFFECTS OF H2O2 AND SNP ON *T. THERMOPHILA*

Exposure of *T. thermophila* to the IC50 of H2O2 or SNP induced a significant increase in the intracellular activities of catalase, SOD and the level of lipid peroxides (substances reactive with thiobarbituric acid (TBARS) or malondialdehyde (MDA)) compared with control (Table 5). Cells treated with SNP showed the greatest increase in catalase activity with a 2.5-fold increase compared to control, while the catalase activity of cells treated with H2O2 increased significantly by 2.2-fold. Similarly, a 4.8-fold increase in SOD activity was observed in SNP-treated cells, while in H2O2-treated cells it increased by 4.5-fold. A clear increase in TBARS was observed in both stress conditions. The level of lipid peroxidation measured by TBARS was increased 3.4- and 1.6-fold in SNP- and H2O2-treated cells, respectively.

Table 5. *In vivo* **effects of H2O2 and SNP on antioxidant markers and GAPDH in** *T. thermophila*

	Timer	H2O2	SNP
Catalase *(umol min mg protein)*	$11,317 \pm 5,054$	$24,867 \pm 4,999$**	$28,019 \pm 1,918$**
Superoxide dismutase *(umol/min/mg protein)*	$0,010 \pm 0,003$	$0,048 \pm 0,008$*	$0,051 \pm 0,017$*
Substances reactive with thiobarbituric acid *(nmol/mg protein)*	$0,039 \pm 0,001$	$0,059 \pm 0,008$*	$0,131 \pm 0,002$**
GAPDH *(umol/min/mg protein)*	$2,010 \pm 0,103$	$1,530 \pm 0,033$**	$0,620 \pm 0,019$**

The values represent the averages of 3 separate experiments ± standard deviations.

1   Significantly different from the control value at p < 0.05

*2   * Significantly different from the control value ap < 0.01*

In order to elucidate the *in vivo* effect of oxidative and nitrosative stress on *T. thermophila* GAPDH, we determined the activity of GAPDH in crude extracts obtained from *T. thermophila* cells cultured in the presence of both stress agents (H2O2 or SNP) at IC50. The results show that GAPDH is highly sensitive to H2O2 and SNP. Both stress agents significantly reduced the specific activity of GAPDH compared with the control. Thus, the specific activity of GAPDH in SNP-treated cells ёla11 decreased by 69%, whereas in H2O2-treated cells it was decreased by only 24%.

In addition, crude extracts obtained from cells cultured for 72 h at 32°C in the presence of stress agents (H2O2 or SNP) at IC50 as well as those obtained from control cells were analysed by western-blot. A rabbit polyclonal antibody, which we have previously produced against GAPDH purified from *T. thermophila* (Errafiy and Soukri, 2012), was used for the analysis. The results presented in Figure 14 revealed a clear increase in GAPDH expression in the crude extract of cells cultured in the presence of SNP, whereas only a Ыдёге increase was observed when cells were cultured in the presence of H2O2.

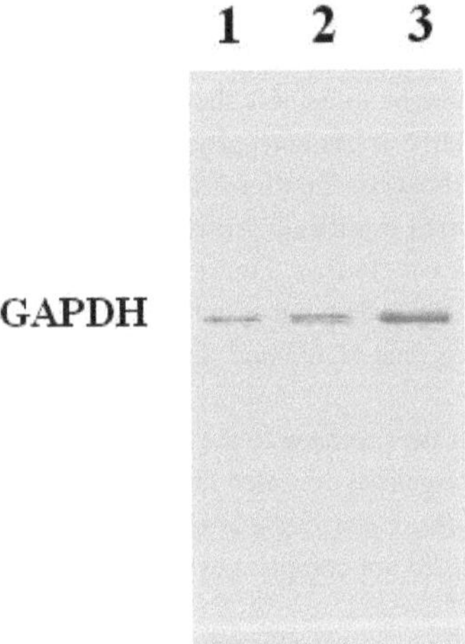

Figure 14. Western blot analysis of GAPDH in crude extracts from *T. thermophila* cells cultured at 32°C for 72 h in the absence of stress agents (1) and (2).

in the presence of H2O2 (2) and SNP (3) at FIC50

A polyclonal antibody produced against purified GAPDH from *T. thermophila* was used.

IV. *IN VITRO* EFFECT OF STRESSORS ON GAPDH ACTIVITY

In order to study the *in vitro* effect of oxidative and nitrosative stress on *T. thermophila* GAPDH, the enzyme was purified to electrophoretic homogeneity from a crude cell extract using a procedure involving ammonium sulphate precipitation and two chromatographic steps, DEAE-cellulose and mono-S (Errafiy and Soukri, 2012). The purified enzyme was incubated in the presence of different concentrations of H2O2 or SNP for 30 min a

30°C. The results show that both types of stress reduce GAPDH activity. H2O2 and SNP have a dose-dependent inhibitory effect on GAPDH activity; as the concentration of the stress agent increases, GAPDH activity decreases until complete inhibition at 25 mM H2O2 and 75 mM SNP. Determination of the IC50 by probit analysis (Bliss, 1935) indicated that of the two stress agents studied, H2O2 was the more toxic (0.853 ± 0.048 mM) compared with SNP (4.115 ± 0.091 mM) (Table 6).

Table 6. In *vitro* inhibitory concentrations (IC10, IC50 and MIC) of H2O2 and SNP on the GAPDH activity of *T. thermophila*

	IC10 (mM)	IC50 (mM)	MIC (mM)
H2O2	0,096 ± 0,015	0,853 ± 0,048	25
SNP	0,233 ± 0,027	4,115 ± 0,091	75

The values represent the averages of 3 separate experiments ± standard deviations.

V. DISCUSSION

In this study, we demonstrated that exposure of *T. thermophila* to stress agents causes oxidative damage to a number of parameters. Two different stress agents (H2O2 and SNP) were used to assess the toxic effects on the protozoan. The effects of nitrosative stress on micro-organisms have already been studied by several researchers using SNP as a nitrosating agent (Rogstam et *al.*, 2007). The effect of H2O2 and SNP on the growth of *T. thermophila* showed that the two stress agents acted differently, but that both completely inhibited growth (Figure 12). Similar results were observed in the ciliate *Tetrahymena pyriformis* (Fourrat et al., 2007b). H2O2 is also thought to inhibit the growth of the yeast *Yarrowia lipolytica* (Biriukova et *al.*, 2006). Similarly, it has been observed in human lung fibroblasts that H2O2 induced growth arrest at sublethal concentrations (Lee et *al.*, 2000). With regard to nitrosative stress, Sahoo et *al* (2003) reported significant growth inhibition in the presence of GSNO (NO donor compound) in certain yeasts. Indeed, Figure 12 and Table 3 reveal that H2O2 is more toxic than SNP for our cell model. The number and time of generation were also modified by the two stress agents, indicating that H2O2 and SNP acted significantly on the growth of *T. thermophila*.

On the other hand, exposure of the protozoa to the IC50 of H2O2 or SNP strongly modified cell morphology. Microscopic observations revealed that SNP induced significant changes in the cell structure of *T. thermophila*. Cells treated with SNP displayed unusual shapes. Their motility was affected and they did not complete cytokinesis. Similar effects were observed in *T. pyriformis*.

^aᴎёe to ethidium bromide (Meyer et *al.*, 1972). H2O2 also affected the protozoan cell structure by increasing its internal volume. These effects were reversible after transfer of the stressed cells to a culture medium without stress agents.

Certain biomarkers can provide information on the state of intracellular stress levels. To counter oxidative and nitrosative stress, cells use a wide range of enzymatic and non-enzymatic defence mechanisms (Powers and Jackson, 2008). Aerobic cells have developed defence strategies against damage induced by oxidising species, including defensive enzymes such as catalase and SOD (Halliwell and Gutteridge, 1999). In our study, the presence of stressors induced a significant increase in the antioxidant defence system (Table 5) of *T. thermophila*. Catalase activity increased by a factor of 2.5 and 2.2 respectively in the presence of SNP and H2O2. This increase in catalase activity may reflect a compensatory mechanism aimed at attenuating the oxidative and nitrosative pathways. It has been suggested that catalase is an important pathway for the decomposition of H2O2 (Coien and Hocistein, 1963), which must be eliminated rapidly because of its toxicity (Halliwell and Gutteridge, 1999). A significant increase in SOD activity was observed in *T. thermophila* cells treated with SNP and H2O2 compared with control cells. The activities were 4.8 and 4.5 times higher than those measured in the control culture. The increase in SOD activity probably reflects an increase in the role of the SOD system in defence against oxidative and nitrosative stress. SOD catalyses the dismutation of superoxide anions into hydrogen peroxide, and catalase then converts H2O2 into molecular oxygen and water (Inal et *al.*, 2001). Table 5 shows a significant increase in lipid peroxidation in *T. thermophila* cells treated with SNP and H2O2. The level of lipid peroxides was higher in the presence of SNP in the culture medium than in the presence of H2O2, with a factor of 3.4 for SNP compared with only 1.6 for H2O2. The increased levels of TBARS suggest a clear increase in free radical levels, which could be due to their increased production and/or reduced destruction (Giugliano et *al.*, 1996). Lipid peroxidation is a self-catalytic destructive process induced by free radicals, whereby polyunsaturated fatty acids in cell membranes undergo degradation to form lipid iodine peroxides (Slater, 1984; Sevanian and Hocistein, 1985). These compounds decompose to form a wide variety of products, in particular MDA (Zeyuan et *al.*, 1998).

As the stress agents used had a strong influence on the growth and morphology of *T. thermophila*, their physiological effects have been extensively studied. GAPDH was therefore chosen as a metabolic marker. Analysis of GAPDH in the crude extract of *T. thermophila* after treatment with H2O2 or SNP revealed a decrease in the specific activity of this enzyme. A significant decrease was observed in *T. thermophila* cells treated with SNP, resulting in a 69% loss of GAPDH specific activity, while in cells treated with H2O2, the decrease was estimated at 24%. A similar study carried out in our laboratory by Fourrat et *al* (2007b) on *T. pyriformis* showed that H2O2 had no effect on the specific activity of GAPDH, whereas SNP reduced it. On the other hand, to assess the level of GAPDH expression in

T. thermophila exposedëes to stress agents, 1 western blot analysis using a rabbit polyclonal antibody produced against the purified enzyme was eiTected. The results showed a clear increase in GAPDH expression in cells cultured in the presence of the SNP, while those cultured in the presence of H2O2 showed only a slight increase. The increase in the level of GAPDH protein expression observed *in vivo* suggests a cellular response to compensate for the inhibitory effect on activity observed in both cases of stress (Fourrat et *al.*, 2007b).

The *in vitro* effect of stress agents on the activity of GAPDH purified from *T. thermophila* was also evaluated. The native enzyme was purified from cells grown in culture medium without stressors using ammonium suliate precipitation followed by two chromatographic steps, DEAE-cellulose and mono-S (Erraiiy and Souki, 2012). The native enzyme was incubated for 30 min at 30°C in the presence of different concentrations of H2O2 or SNP to determine the impact of *in vitro* exposure on GAPDH activity. The results showed that increasing the concentration of the stressor was accompanied by a decrease in GAPDH activity. Complete inhibition of GAPDH activity was observed at 25 mM H2O2 and 75 mM SNP. Furthermore, determination of the IC50 indicates that GAPDH is much more sensitive to H2O2 than to SNP, since the IC50 for H2O2 is approximately 0.853 ± 0.048 mM, whereas for SNP it is 4.115 ± 0.091 mM. These results could be explained in part by the presence of a highly reactive thiol in the GAPDH active site, which can be affected by free radicals. Indeed, it has been reported that inactivation of GAPDH is mainly caused by radical interference with the cysteine residue present in the active site, which has been described as S-thiol by H2O2 (Schuppe-Koistinen et *al.*, 1994) and S-nitrosyl by nitric oxide (Stamler et *al.*, 1992). Grant et *al* (1999) demonstrated that the activity of GAPDH is regulated by S-thiolation of the protein in protection against oxidative stress and that this process is physiologically important for survival under conditions of oxidative stress. GAPDH has been shown to undergo endogenous ADP-ribosylation in response to SNP (Kots et *al.*, 1992). It has also been described that nitric oxide radicals can lead to inactivation of GAPDH by promoting modification of the cysteine residue of the active site by mono-ADP-ribosylation (Kots et *al.*, 1992; Zhang and Snyder, 1992). Furthermore, it has been reported that proteins crucial for cellular metabolism or structure, such as GAPDH, can be mono-ADP-ribosylated by oxidative stress (Dimmeler et *al.*, 1992).

VI. CONCLUSION

A better understanding of the damage induced by stress agents affecting the cell may guide future research towards therapies to combat oxidative and/or nitrosative stress. One interesting avenue of research is to stimulate endogenous systems using natural products. The main objective of this study is to elucidate the consequences caused by stress on the growth, morphology and physiology of the cilius *T. thermophila,* considered as a eukaryotic cell model that can simulate the short-term effects on human cells. Our results indicate that H2O2 and SNP dë trigger changes in cell growth and morphology and induce *in vivo* and *in vitro* inhibition of GAPDH activity.

PART 2: PURIFICATION AND CHARACTERISATION OF THE GLYCERALDEHYDE-3-PHOSPHATE DESHYDROGENASE OF THE CILY PROTOZOAR *Tetrahymena thermophila*

In order to assess the metabolic perturbations caused by oxidative/nitrosative stress on the ciliated protozoan *T. thermophila* used as a cellular model, we focused, in this second part, on the study of the *in vitro* effect of two stress agents (H2O2 and SNP) on glyceraldehyde-3-phosphate dehydrogenase (GAPDH), chosen as the enzymatic model. To this end, we purified this key carbohydrate metabolism enzyme from crude *T. thermophila* cell extract using a new procedure involving fractional precipitation with ammonium sulphate followed by two column chromatography steps (DEAE-cellulose: anion exchange and Mono-S: cation exchange).

I. PURIFICATION OF GAPDH FROM *T. THERMOPHILA*

The results obtained show that the crude extract of *T. thermophila* cells lysed by sonication provides a total protein content of 986 mg, corresponding approximately to 642 units of GAPDH (Table 7). After fractional precipitation with ammonium sulphate in the range 55-88%, eliminating 92% of contaminating proteins, the dialysed enzyme fraction was introduced into the DEAE-cellulose column and elution was carried out at a constant flow rate of 12 ml/h. This step yielded 380.61 units of GAPDH in 76.23 mg of protein (Table 7). At this stage the enzyme is partially purified. The recovered enzyme solution was then introduced into the Mono-S column. This was thoroughly washed to remove contaminating proteins. The GAPDH was then eluted using a linear gradient of KCl (from 0 to 300 mM) at a flow rate of 12 ml/h. This final purification step enabled us to obtain a homogeneously pure protein with a specific activity of 44 U/mg of protein for a yield of approximately 8% and a purification factor of around 68-fold (Table 7).

Table 7. Purification of GAPDH from *Tetrahymena thermophila*

Fraction	Total protein (mg)	Specific activity (U/mg protein)	Total activity (U)	Purification factor (times)	Yield (%)
Raw extract	986,06	0,651	642,02	1,00	100,00
Ammonium sulphate (55-88%)	76,23	4,992	380,61	7,66	59,28
DEAE-cellulose	12,15	9,197	111,75	14,12	17,40
Mono-S	1,20	44,500	53,40	68,34	8,31

To verify Гьотодёпёкё and the purity of GAPDH, we analysed the different fractions recovered during purification by denaturing medium polyacrylamide gel electrophoresis (SDS-PAGE). The results obtained show a significant reduction in the number of contaminating protein bands after each purification step, in favour of a progressive enrichment of a 32 kDa protein (Figure 15a). This protein corresponds to the putative subunit of *T. thermophila* GAPDH with an estimated molecular weight of 32 kDa.

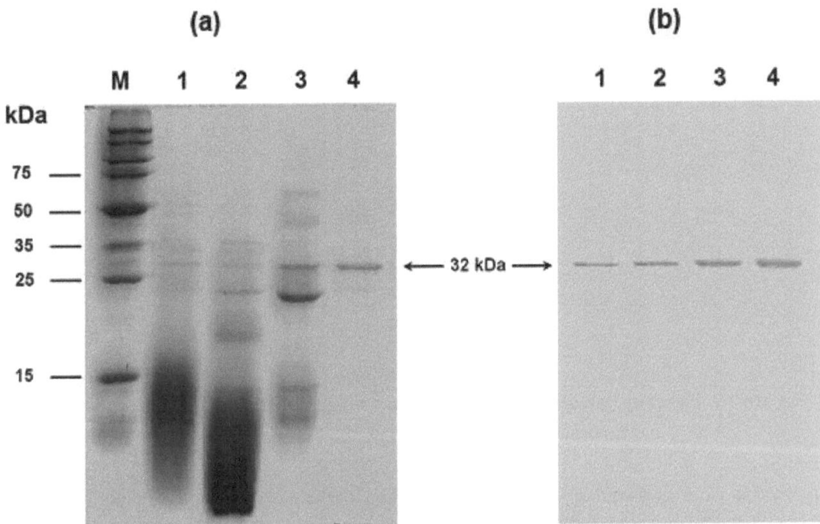

Figure 15. Purification of GAPDH from *Tetrahymena thermophila*

(a) Analysis by denaturing medium polyacrylamide gel electrophoresis (SDS-PAGE) coloured with coomassie blue showing the protein profiles of the different fractions obtained obtained during purification. Line M, Markers; Line 1, Crude extract; Line 2, Ammonium sulphate fraction
55-88%; Line 3, DEAE-cellulose fraction; Line 4, Mono-S fraction.
A similar amount of protein was applied to each line.
(b) Western blot analysis using polyclonal antibodies specific for GAPDH from *T. thermophila*. Line 1, crude extract; Line 2, 55-88% ammonium sulphate fraction; Line 3, DEAE-cellulose fraction; Line 4, Mono-S fraction. The arrow indicates the band of the 32 kDa GAPDH subunit.

II. WESTERN BLOT ANALYSIS

Rabbit polyclonal antibodies were raised against purified GAPDH from *T. thermophila*. These antibodies reacted selectively by immunoblotting with a single immunoreactive band in both the crude extract and the purified preparations. Figure 15b shows the reaction of GAPDH antibodies which clearly recognised a single 32 kDa band, corresponding to the GAPDH subunit. No protein bands were detected using pre-immune serum as the primary antibody.

III. DETERMINATION OF MOLECULAR WEIGHT

To determine the molecular weight of the native enzyme, gel filtration by FPLC was performed and gave a value of approximately 120 kDa corresponding to the molecular weight of native *T. thermophila* GAPDH (Figure 16). This result, compared with that obtained by SDS-PAGE which showed a single band corresponding to the 32 kDa protein, suggests that purified GAPDH from *T. thermophila* should have a homotetrameric structure.

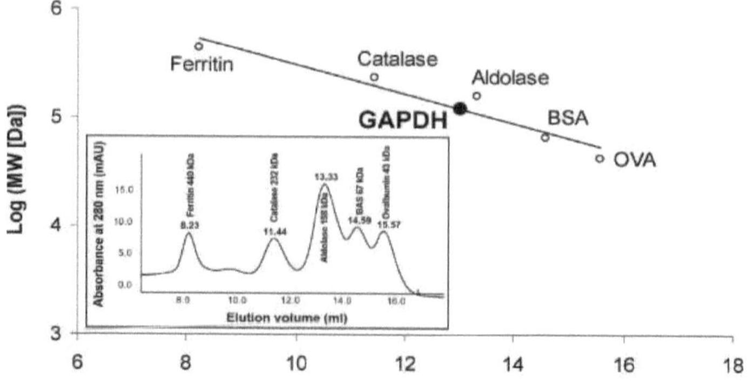

Figure 16. Determination of the molecular weight of native GAPDH from *T. thermophila* on a Superdex 200 HR column

The proteins were applied to a Superdex column. The markers used to calibrate the calibrate the column and establish a standard curve were ferritin (440 kDa), catalase (232 kDa), aldolase (158 kDa), BSA (67 kDa) and ovalbumin (43 kDa). The elution of proteins was monitored by spectrophotometry at 280 nm.

IV. *pI* DETERMINATION

Protein isoëlectrofocusing analysis based on *pI* values showed a single protëine band at 8.8 (the estimated pÏë for the enzyme) (Figure 17). This result indicates that there is only one basic isoform of the purified enzyme, which strongly suggests the expression of a single copy of the *gapC* gene.

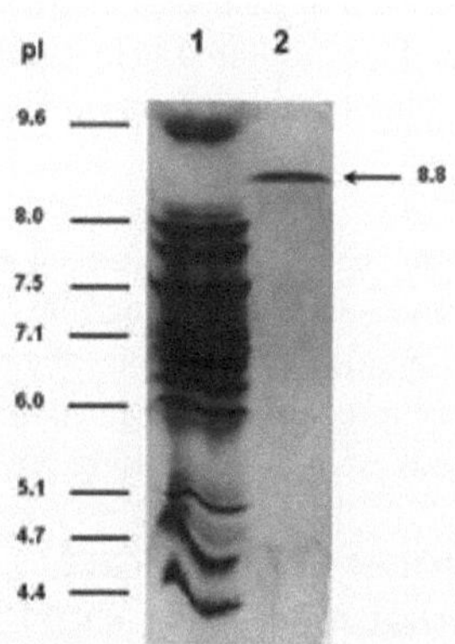

Figure 17. Isoelectrofocusing analysis of GAPDH from *T. thermophila*

Isoëlectrofocusing is performed in polyacrylamide gel (5% w/v, acrylamide) with a pH gradient generated by ampholites (pH range 3.5-10). 1, Markers; 2, GAPDH purified from *T. thermophila* (30 |ig). The arrow indicates the isoelectric point *(pI) of* GAPDH.

V. INFLUENCE OF pH AND TEMPERATURE ON PURIFIED GAPDH ACTIVITY

The activity profile of purified GAPDH was determined over a pH range of 4.0 to 9.5 using a mixture of different buffers. The enzyme exhibits a typical bell-shaped profile over a wide pH range (Figure 18a). Enzyme activity is maximal at a pH of about 8.0.

The influence of temperature on enzyme activity was determined between 5 and 65°C. Studies of the effect of temperature on enzymatic activity (thermal activation) revealed an optimum temperature of around 30-35°C (Figure 18b). Preincubation of *T. thermophila* GAPDH for 10 minutes at temperatures ranging from 5 to 40°C (thermal denaturation) did not irreversibly affect enzyme activity. However, thermal inactivation occurred above 40°C. Increasing the temperature above 40°C does not increase the enzyme's kinetic energy but disrupts the forces that maintain the molecule's shape; the enzyme is progressively denatured, leading to changes in the shape of the active site. Temperatures above 60°C completely denature the enzyme.

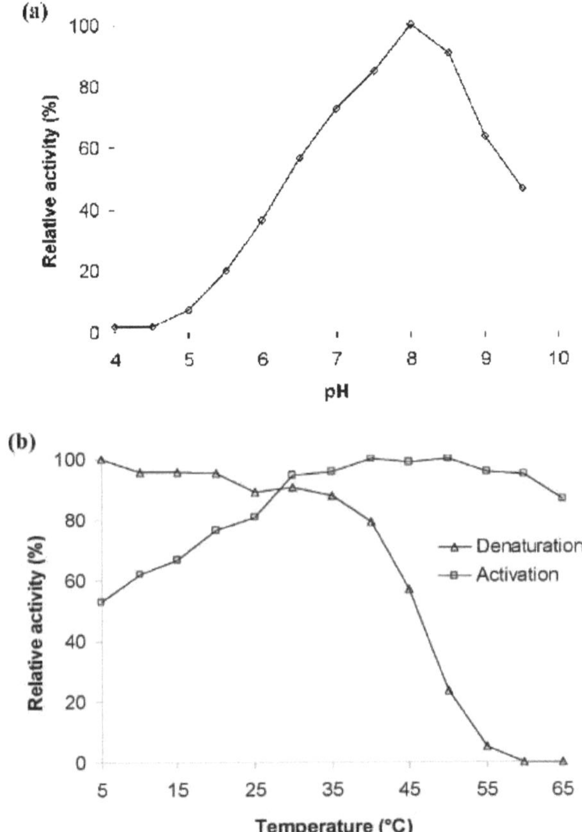

Figure 18. Effects of pH and temperature on the enzymatic activity of *T. thermophila* GAPDH

(a) The enzymatic activity of *T. thermophila* GAPDH in the pH range 4.0-9.5 using a mixture of different buffers.

(b) GAPDH enzymatic activity at temperatures between 5 and 65°C; effects of measurement temperature (activation) and

enzyme
pre-incubation temperature (denaturation). The values represent the
averages of 3 separate experiments

VI. CINETIC STUDIES OF PURIFIED GAPDH

The initial rates of the enzymatic reaction were determined by varying the concentration of the substrates, NAD+ (from 0.02 to 0.32 mM) or D-G3P (from 0.02 to 0.64 mM), since GAPDH catalyses a two-substrate reaction. The apparent K_m values for NAD+ and D-G3P were estimated at 102±12 µM and 360±18 µM, respectively. The V_{max} of purified protëine was estimëed at 39.40±2.95 U/mg (Table 8). The K_m NAD+ of *T. thermophila* GAPDH is clearly lower, suggesting a more élevëe affinity for the nucleotide coenzyme.

Table 8. Comparison of some kinetic and physicochemical parameters of GAPDH purified from *T. thermophila* with that of other organisms

(Hafid et *al.*, 1998; Mounaji et *al.*, 2002; Baibai et *al.*, 2007; Mountassif et *al.*, 2009)

Origin of GAPDH	K_m D-G3P (µM)	K_m NAD$^+$ (µM)	V_{max} (U/mg of protein)	Optimum pH	Optimum temperature (°C)
Tetrahymena thermophila	360±18	102±12	39,4±2,9	8,0	30-35
Tetrahymena pyriformis	150±5	5±1	5,6±0,8	8,5	35
Sardina pilchardus	73,4±8,1	92±7,4	37,6±2,9	8,0	28-32
Pleurodeles waltl	27±11	60±10	33,2±5,7	8,5	40
Camelus dromedarius	210±80	25±10	52,7±5,9	7,8	28-32
Human erythrocytes	20,7	17,8	4,3	8,5	40-45

VII. EFFECT OF H2O2 AND SNP ON PURIFIED GAPDH

In order to study the physiological effect of oxidative and nitrosative stress on *T. thermophila*, we chose GAPDH as an enzymatic marker sensitive to H2O2 and NO as demonstrated in previous studies on *T. pyriformis* (Fourrat et *al.*, 2007b) and *Saccharomyces cerevisiae* (Cabiscol et *al.*, 2000) cells. Purified GAPDH from *T. thermophila* was incubated in the presence of different concentrations of two stress agents: H2O2 and SNP.

Figure 19 shows the dose-response relationships on GAPDH activity. Both oxidative and nitrosative stresses induced a decrease in GAPDH activity up to competitive inhibition at 25 and 75 mM respectively. The results also show that 1 mM H2O2 can inhibit GAPDH activity by up to 50% at 30°C after 30 min incubation (Figure 19a), while SNP decreases GAPDH activity by 50% at 5 mM after the same incubation time (Figure 19b). The effect of H2O2 on GAPDH activity showed a standard sigmoidal dose-response curve (Figure 19a), whereas SNP exerted a biphasic inhibitory effect as indicated by the two different slopes of the curve (0-5 mM and 5-75 mM) (Figure 19b). Research is underway to clarify these results.

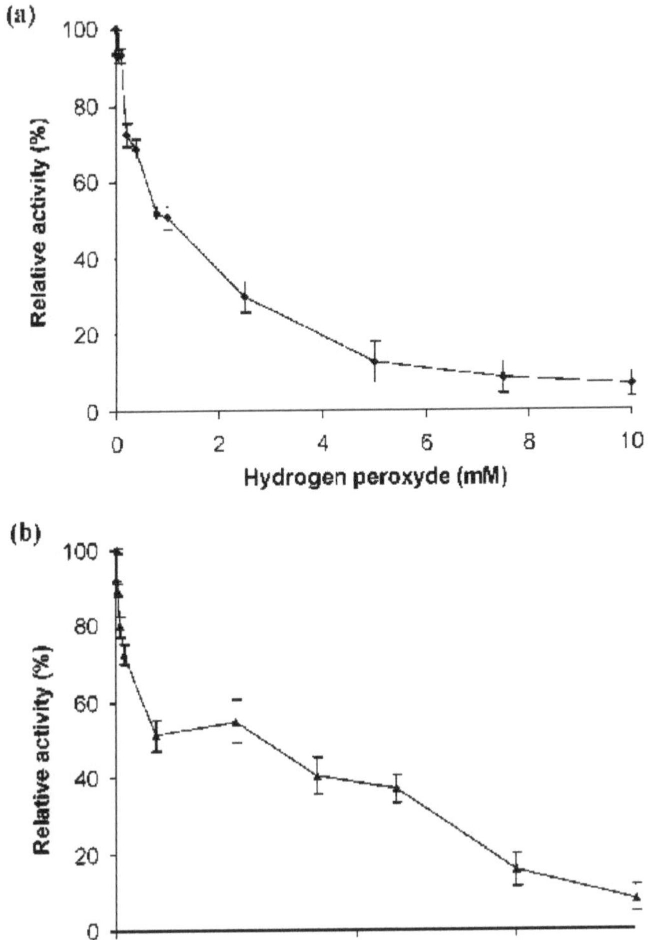

Figure 19. Effect of oxidative and nitrosative stress on GAPDH activity in *T. thermophila.*

(a) Inhibition of GAPDH purifiée by гН2О2. GAPDH activity was measured after 30 min incubation with different concentrations of H2O2. (b) Inhibition of GAPDH purifiée by SNP. GAPDH activity was measured after 30 min incubation with different concentrations of SNP. Values represent averages of 3 expëriences sëparëes.

VIII. DISCUSSION

GAPDH was purified to ëlectrophorëtic liomogeneity from the crude cell extract of the ciliated protozoan *T. thermophila* using a procedure involving fractional precipitation with ammonium sulphate, followed by two ciromatographic steps (anion eicanger and cation ccianger). This procedure was carried out at 4°C to minimise protein degradation. As indicated above, SDS-PAGE and Western blot analysis of the purified enzyme showed a single band corresponding to a 32 kDa protein (Figure 15). This result, compared with the native molecular weight determined by FPLC gel filtration (120 kDa) (Figure 16), suggests that the purified enzyme has an iomotetrameric structure like other GAPDHs (Fortiergill and Miciels, 1993). However, the molecular weight of the purified GAPDH subunit (32,000 Da) is somewhat lower than that estimated from the protein found (36,800 Da) in another strain of the same microorganism (*T. thermophila*) (Zhao et *al.*, 1997). The nucleotide sequence of gëne GAPDH (*gapC*) from *T. thermophila* is listed in GenBank under accession number AF319450.1. The difference observed between the cstimated molecular weight and the theoretical molecular weight may be the result of several types of post-translational events, such as alternative splicing (AS), endoproteolytic processing (EPP) and post-translational modifications (PTM) (Ahmad et *al.*, 2005).

Analysis of purified GAPDH by the isoelectrofocusing technique revealed a single isoform at 8.8 (the estimated *pI* for the enzyme) (Figure 17). This result strongly suggests that only one copy of gëne *gapC* is expressed. These data are in agreement with those reported by Zhao et *al.* (1997). A single *gapC* isoform has also been found in *T. pyriformis* (Hafid et *al.*, 1998) and other eukaryotic and prokaryotic organisms (Mounaji et *al.*, 2002), but this does not appear to be a general rule since the presence of several GAPDH isoforms has been reported in other organisms (Soukri et *al.*, 1996).

The optimum pH for the enzymatic reaction of purified GAPDH was 8.0 (Figure 18a). An identical value was obtained for GAPDH from *Sardina pilchardus* (Baibai et *al.*, 2007). However, for GAPDH from *T. pyriformis*, the optimum pH was 8.5 (Hafid et *al.*, 1998) (Table 8). On the other hand, studies on the effect of temperature on the activity of *T. thermophila* GAPDH revealed an optimum value between 30-35°C (Figure 18b). A similar value was obtained for GAPDH from *T. pyriformis* (Hafid et *al.*, 1998). However, GAPDH from human erythrocytes (Mountassif et *al.*, 2009) and *Pleurodeles waltl* (Mounaji et *al.*, 2002) showed different optimum temperature values (40-45°C and 40°C, respectively) (Table 8).

As GAPDH catalyses a two-substrate reaction, the Michaelis constant values (K_m) and maximum rates (V_{max}) for NAD reduction$^+$ and D-G3P oxidation by purified GAPDH were determined graphically from Lineweaver-Burk double reciprocal diagrams (1934). The K_m values for NAD$^+$ and D-G3P were estimated to be 102 ± 12 and 360 ± 18 µM, respectively. The V_{max} of the purified enzyme was approximately 39.40 ± 2.95 U/mg (Table 8). Consequently, the K_m NAD$^+$ of *T. thermophila* GAPDH is clearly ini'eral, suggesting a higher ai'finity for the pyridine coenzyme, as has been observed in *T. pyriformis* (Hafid et *al.*, 1998) and *Camelus dromedarius* (Fourrat et *al.*, 2007a). For GAPDH purified from *S. pilchardus* and *P. waltl*, the V_{max} (37.6 and 33.24 U/mg, respectively) was similar to that obtained for GAPDH from *T. thermophila*. Also, the K_m NAD$^+$ of *S. pilchardus* GAPDH (Baibai et *al.*, 2007) was similar to that estimated for *T. thermophila* GAPDH, but different from the K_m NAD$^+$ of *P. waltl* GAPDH (Mounaji et *al.*, 2002). In addition, the K_m D-G3P was approximately similar to those found for GAPDH from *T. pyriformis* (Hafid et *al.*, 1998) and *C. dromedarius* (Fourrat et *al.*, 2007a), but different from those reported for GAPDH from *S. pilchardius* (Baibai et *al.*, 2007) and *P. waltl* (Mounaji et *al.*, 2002) (Table 8).

Investigations into the effect of oxidative and nitrosative stress on *T. thermophila* GAPDH show that H2O2 and SNP induce a decrease in GAPDH activity until complete inhibition

(Figure 19). It has been reported that GAPDH inactivation is mainly caused by radical interference with the cysteine residue present in the active site, which has been described as S-thiole by H2O2 (Schuppe et *al.*, 1994) and S-nitrosyl by NO (Stamler et *al.*, 1992). H2O2 has also been shown to modify the cysteine residues of the GAPDH catalytic site, resulting in both inactivation of GAPDH enzymatic activity and structural modifications of GAPDH (Beak et *al.*, 2008). NO inhibits GAPDH activity by modifying thiols, which are essential for this activity, and the modification includes sulphenic acid formation (Ishii et *al.*, 1999). Nitrosylation of Cys-149 in the active site of GAPDH causes a decrease in enzymatic activity (Mohr et *al.*, 1999). We have shown here that 1 mM H2O2 can inhibit GAPDH activity by up to 50% at 30°C after 30 min incubation (Figure 19a), while SNP decreases GAPDH activity by 50% at 5 mM after the same incubation time (Figure 19b). These observations are in agreement with those reported by Fourrat et *al* (2007b) for the same enzyme in *T. pyriformis*, which suggest that GAPDH could be used to characterise the physiological effect of oxidative and nitrosative stress. A study carried out on U937 cells (human promonocytes derived from a histiocytic lymphoma) showed that GAPDH is inactivated by H2O2, as part of a cellular defence pathway against apoptosis induced by oxidative stress (Colussi et *al.*, 2000). The degree of inhibition depends on the nature of the stress, so H2O2 should be more toxic than SNP, since H2O2 acts at lower concentrations.

IX. CONCLUSION

In conclusion, this study is the first, to our knowledge, to report on the purification and characterisation of glyceraldehyde-3-phosphate dehydrogenase from a strain of *T. thermophila*. The procedure used for this purification was rapid and simple, involving two chromatographic steps, namely DEAE-cellulose and Mono-S. The purified *T. thermophila* GAPDH has a homotetrameric structure. This GAPDH differs in several respects from those previously described from other organisms (Hafid et *al.*, 1998; Mounaji et *al.*, 2002; Baibai et *al., 2007*; Fourrat et *al.,* 2007a). A study of the effect of two types of stress (oxidative and nitrosative) on this enzyme showed that oxidative stress is more damaging than nitrosative stress. The physicochemical propriëtës of this GAPDH, ëtant caractërisëes, could be used to ëvaluate the effects of two stress agents (H2O2 and SNP) on the protozoan ^Hë *T. thermophila*, in particular the changes occurring at the cellular and physiological level.

PART 3: PROTECTIVE EFFECT OF CERTAIN ESSENTIAL OILS

ON *TETRAHYMENA THERMOPHILA* EXPOSED TO OXIDIZING AND NITROSATING STRESSES

In this section, we studied the protective effect of certain essential oils against oxidative and nitrosative stress. To evaluate this effect, we monitored and compared the growth kinetics of *T. thermophila* grown for 140 h at 32°C under stressful conditions in the absence and presence of essential oils.

These essential oils have been extracted from 7 plants known for their therapeutic properties. They were kindly supplied by Professor Emna Ammar (*Sfax National Engineering School*) and stored at 4°C until use.

These oils were extracted from lavender (*Lavandula angustifolia*), geranium (*Pelargonium robertianum*), thyme (*Thymus capitalus*), rosemary (*Rosmarinus officinalis*), cypress (*Cupressus sempervireus*), juniper (*Juniperus phoenica*) and clove (*Syzygium aromaticum*) (Table 9).

I. GROWTH RESPONSE OF *T. THERMOPHILA* UNDER STRESS CONDITIONS

H_2O_2 and SNP were used to create oxidative and nitrosative stress on the protozoan, respectively. Both stress agents completely inhibited the growth of *T. thermophila* at concentrations of 1 mM H_2O_2 and 10 mM SNP (Figure 20).

Thus, the IC_{50}s of H_2O_2 (0.7 mM) and SNP (1.8 mM), estimated using probit analysis from Figure 20, were used to assess the effect of these stressors on the growth of *T. thermophila*.

The typical growth curve (lag, exponential and stationary phases) was significantly modified by the addition of H_2O_2 or SNP. Thus, a significant decrease in the lag phase was observed in cultures supplemented with stress agents compared with the control (Figure 21). Similarly, the exponential phase showed a significant decrease in growth under oxidative and nitrosative stress. No change was observed in the stationary phase under oxidative stress compared with the control. However, a significant change in the stationary phase was observed under nitrosative stress compared to control (Figure 21).

Table 9. Major components of essential oils tested against oxidative and nitrosative and nitrosative stress

Species	Oil detail		
	Common name	**Majority compound (area, %)***	**Chemical structure of the majority compound**
Lavandula angustifolia	Lavender	Linalyl acetal (34.5), Linalool (20.5), Methylcyclopentane (10.2)	
Pelargonium robertianum	Geranium	Geranyl acetate (42.3), Geraniol (36.7)	

Thymus capitalus	Thyme	Borneol (24.1), Thymol (6.7)	
Rosmarinus officinalis	Rosemary	1,8-cineole (27.22), Trans-β-ocimene (10.7), Camphor (10.4)	
Cupressus sempervirens	Cypress	δ-3-carene (22.9), α-pinene (20)	
Juniperus phoenica	Juniper	α-pinene (59.1), Linalool (3.3)	
Syzygium aromaticum	Cloves	Eugenol (75.3), Eugenyl acetate (15.1)	

[*] According to GC-MS analysis (Ben Saida, 2007).

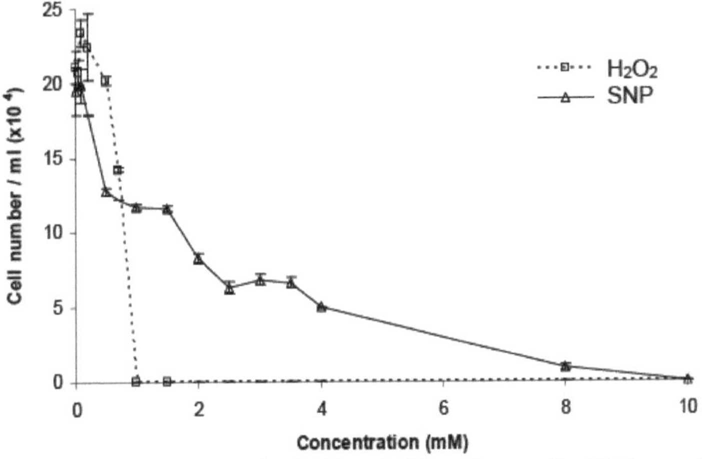

Figure 20: Dose-response curves for H2O2 and SNP on the growth of *T. thermophila*. The number of cells has ële determined after 72 h of growth at 32°C in PPYE medium containing different concentrations of stress agent (H2O2 or SNP).

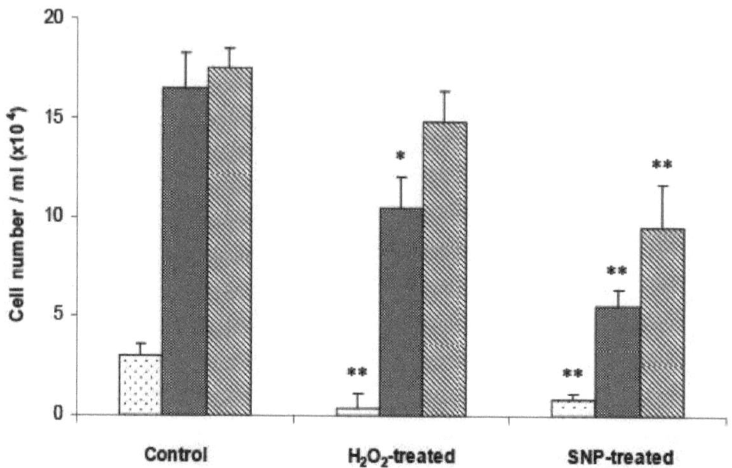

Figure 21. Effects of H2O2 and SNP on the growth of *T. thermophila*

Cell numbers were determined after 24, 72 and 140 h of growth at 32°C, corresponding to the latent, exponential and stationary phases respectively.

T. thermophila was grown in PPYE medium containing 0.7 mM H2O2 or 1.8 mM SNP. *T. thermophila* grown in the absence of stressors was used as a control. For comparisons, Student's *t-test* was used (* $p < 0.05$ and ** $p < 0.01$).

II. DETERMINATION OF THE SENSITIVITY OF *T. THERMOPHILA* TO ESSENTIAL OILS

The 7 natural essential oils used in this study, in their pure state, inhibited the growth of *T. thermophila* when added to the culture medium. A series of dilutions was therefore carried out to determine the minimum inhibitory concentrations (MICs). The MIC was obtained at dilution 10^{-5}. However, from dilution 10^{-6}, normal growth of *T. thermophila* was observed. For subsequent studies on the anti-stress effect of essential oils, the 10 dilution^{-9} was chosen. Monitoring of the growth curve of *T. thermophila* in the presence of different essential oils at this dilution showed no effect on either growth or cell shape.

III. ANTI-OXIDATIVE STRESS EFFECT OF ESSENTIAL OILS

The essential oils tested (lavender, geranium, thyme, rosemary, cypress, juniper and clove) showed no significant change ($p < 0.05$) in growth at the latent phase compared with the control (*T. thermophila* treated with H2O2). In contrast to the lag phase, significant changes were observed in the exponential phase. The number of *T. thermophila* cells treated with H2O2 was significantly higher ($p < 0.01$) when the cultures were supplemented with cypress or juniper essential oils (Figure 22). Similarly, thyme and rosemary essential oils significantly ($p < 0.05$) increased the number of cells in the exponential phase, whereas no significant change was observed when lavender, geranium or clove essential oils were added to H2O2-treated cultures. The number of cells in stationary phase increased significantly ($p < 0.01$) in the presence of cypress or juniper essential oils compared with the control. Similarly, rosemary essential oil significantly ($p < 0.05$) increased the number of cells in the stationary phase. As with the exponential phase, no significant changes were observed in the stationary phase in the presence of lavender, geranium or clove essential oils. However, thyme essential oil did not significantly affect growth in the stationary phase (Figure 22).

IV. NITROSATIVE ANTI-STRESS EFFECT OF ESSENTIAL OILS

The action of the natural essential oils used in this study on SNP-treated *T. thermophila* is shown in Figure 23. As can be seen, no significant change ($p < 0.05$) in cell number was observed at the latent phase for the different essential oils used compared to the control without essential oils. However, cypress, rosemary, clove and juniper essential oils, added to SNP-treated *T. thermophila* cultures, significantly ($p < 0.05$) increased cell numbers in the exponential growth phase compared with the control, whereas lavender, geranium and thyme essential oils were not sufficiently effective against nitrosative stress. The stationary phase did not change significantly for the different essential oils.

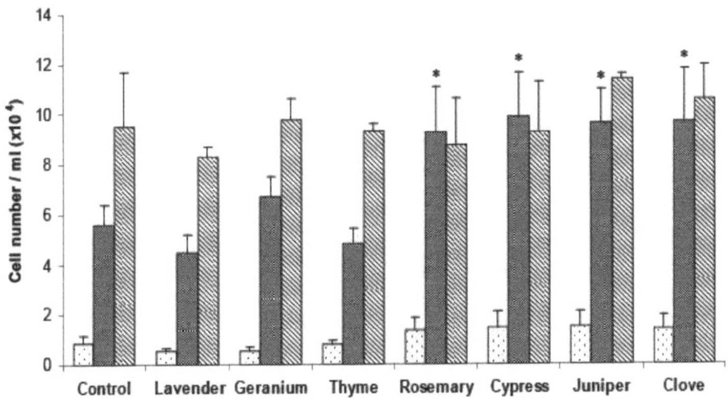

Figure 22. Protective effect of lavender, geranium, thyme, rosemary, cypress, juniper and clove essential oils against oxidative stress caused by 0.7 mM H2O2 on the growth of *T. thermophila*.

The number of cells a ëlë dëtermmë after 24, 72 and 140 h of growth at 32°C, corresponding to the latent, exponential and stationary phases, respectively. For comparisons, Student's *t-test* has ëlë ШШзë (* $p < 0.05$ and ** $p < 0.01$).

The number of cells a ёlё dёterminё after 24, 72 and 140 h of growth at 32°C, corresponding to the latent, exponential and stationary phases, respectively. For comparisons, Student's *t-test* has ёlё ШШзё (* $p < 0.05$ and ** $p < 0.01$).

V. SYNERGISTIC EFFECT OF ESSENTIAL OILS

In order to study whether the тёlапде of essential oils can influence their ability to inhibit the action of stressors, lavender, gёranium and thyme essential oils were ёlё chosen. These essential oils used sёparёment had not shown any interesting effect against the two types of stress (oxidativе and nitrosative).

V .1. Oxidative stress

The addition of two essential oils to H2O2-treated *T. thermophila* cultures showed no significant change ($p < 0.05$) at the lag phase, whereas at the exponential phase, cell numbers increased significantly ($p < 0.01$) with the different essential oil mixtures. Thus, the mixture of lavender and thyme essential oils significantly ($p < 0.01$) increased the number of cells at the exponential and stationary phases (Figure 24). Nevertheless, a significant increase ($p <$ 0.01) in cell number was observed at the exponential phase when thyme and geranium essential oils were mixed. However, no significant changes were observed in the stationary phase (Figure 24). The mixture of lavender and geranium essential oils significantly ($p <$ 0.01) increased the number of cells in the exponential phase with no change in the stationary phase.

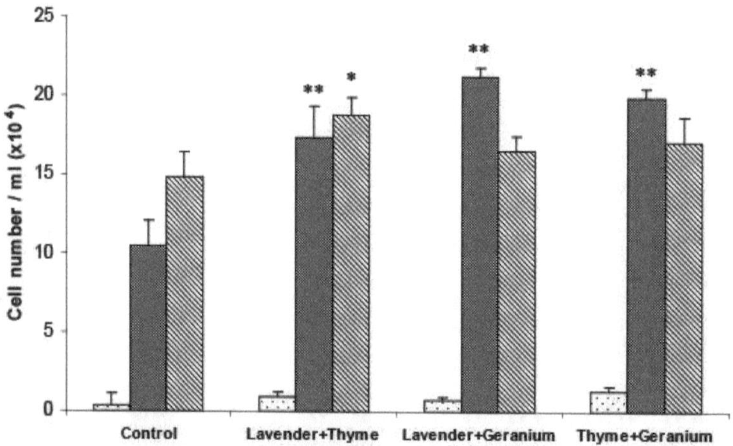

Figure 24. Synergistic effect of 3 essential oils (lavender, geranium and thyme) against the effect of oxidative stress on the growth of *T. thermophila.*

In each experiment, 2 essential oils were added to the culture medium containing 0.7 mM H2O2. Cell numbers were determined after 24, 72 and 140 h of growth at 32°C, corresponding to the latent, exponential and stationary phases respectively. For comparisons, Student's *t-test* was used (* $p < 0.05$ and ** $p < 0.01$).

VI 2. On nitrosative stress

Under nitrosative stress, the synergistic effect of 2 essential oils mëlangëes showed a атёНогайоп at the exponential phase of growth with no change at the lag phase and stationary phase compared to the tëmoin (*T. thermophila* tra^e by SNP) (Figure 25). Indeed, mël mixtures of lavender and thyme, thyme and gëranium, and lavender and gëranium essential oils significantly (p < 0.05) increased cell number at the exponential phase, whereas no significant increase (p < 0.05) in cell number was ëlë observed at the lag phase and stationary phase.

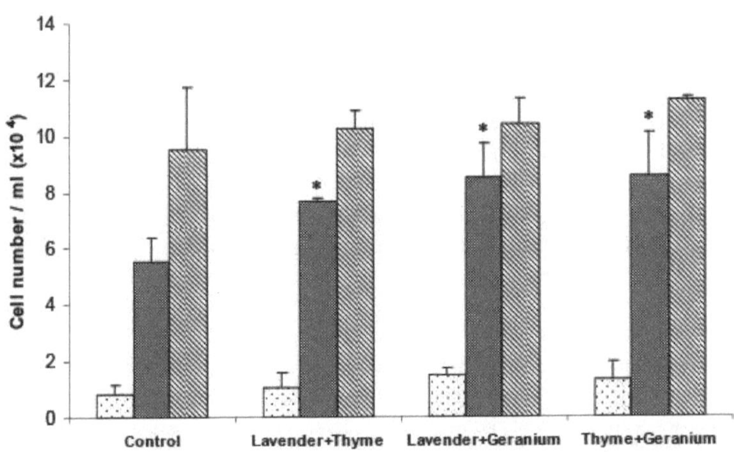

Figure 25. Synergistic effect of 3 essential oils (lavender, geranium and thyme) against the effect of nitrosative stress on the growth of *T. thermophila.*
In each expërience, 2 essential oils were ëlë addedëes to the culture medium containing 1.8 mM of SNP. The number of cells ëlë dëterminë after 24, 72 and 140 h of growth at 32°C, corresponding to the latent, exponential and stationary phases respectively. For comparisons, Student's *t-test* has ëlë utilisë (* p < 0.05 and ** p < 0.01).

VII DISCUSSION

Under normal conditions, *T. thermophila* has a typical growth curve with a lag phase, an exponential phase and a stationary phase. This growth curve is disrupted under stress conditions. The results reported here indicate that PH2O2 and SNP significantly affect the growth curve of *T. thermophila*. As in the case of *Tetrahymena pyriformis* (Fourrat et *al.*, 2007b), the two stress agents (H2O2 and SNP) complemented each other in inhibiting the growth of *T. thermophila* (Figure 20). This could be explained by a disruption in the balance between ROS generation and elimination, which may have consequences for cell survival. Oxidative stress caused a significant decrease in cell number in the first two growth phases (Figure 21). Indeed, a strong toxicity was observed during the latent phase followed by the exponential phase. However, no significant change in cell numbers was observed in the stationary phase. Similar results have been reported for *Yarrowia lipolytica* (Biriukova et *al.*, 2006). Furthermore, Lee et *al* (2000) showed that ГH2O2 at subblethal concentrations induced growth arrest in human lung fibroblasts. Similarly, nitrosative stress led to a reduction in cell numbers in the latent, exponential and stationary phases (Figure 21). The toxic effect of nitrosative stress has already been reported on yeasts by Sahoo et *al* (2003), who observed significant growth inhibition in *Rhodotorula mucilginosa* and *Saccharomyces cerevisiae* in the presence of S-nitrosoglutathione, an NO-donating compound.

Natural essential oils have been used to counter the toxic effects of stressors, and are widely used and well-known for their therapeutic properties. They are used in aromatherapy, cosmetics, to flavour foods and drinks, and also as medicines for cerebral stimulation, anxiolytic sedation and as antidepressants (Buchbauer and Jirovetz, 1994). Due to their concentrated nature, the use of essential oils as natural anti-stress agents requires the establishment of optimal conditions, such as protozoan sensitivity and optimal concentrations. Indeed, previous studies have reported the anti-protozoal effect of essential oils (Mojica et *al.*, 2004). This work is in line with our results, which showed inhibition of the growth of *T. thermophila* in the presence of the various essential oils used in their pure state. However, the essential oils tested at the non-lethal dose of 10^{-9} did not affect the viability or morphology of *T. thermophila cells* when observed under the microscope.

Essential oils have been extensively studied for their antimicrobial, insecticidal, antifungal, antibacterial and cytotoxic activities (Faleiro et *al.*, 1999). The aim of our work was to study the protective potential of natural essential oils to minimise the effects of oxidative and nitrosative stress on *T. thermophila*. The different essential oils tested act differently in relation to stress. Their action differs according to the nature of the plants from which they are derived and the type of stress to which the cell is subjected (oxidative or nitrosative). Throughout this study, the various essential oils showed no anti-oxidative or anti-nitrosative effect during the latency phase. However, interesting results in terms of a significant increase in cell number were observed during the exponential phase. Indeed, cypress and juniper essential oils showed a strong anti-H2O2 effect by significantly increasing cell numbers, while thyme and rosemary showed a moderate protective effect (Figure 22). Similar results were observed in *T. thermophila* treated with SNP, with the exception of thyme, which showed no significant change (Figure 23). The difference observed in the response of *T. thermophila* cells to essential oils depending on their growth stage could be explained by their physiological state, which may affect their adaptive responses to environmental conditions.

On the other hand, it is potentially interesting to correlate the anti-stress behaviour of essential oils with their composition and the chemical nature of their constituents. Thus, as shown in Figures 22 and 23, the most ёкуё anti-stress effect of cypress and genevner essential oils can be attributed to their main constituents, which are 6-3-carene and a-pinene, respectively. Indeed, these composés seem to present a structural analogy compared to the other major composés of the essential oils usedées (Table 9). These essential oils have a different composition, and it is however difficult to attribute the anti-stress effect to one compound. It has been noted that essential oils present a specificity in the amplitude but not in the mode of action of the biological effect (Bakkali et *al.*, 2005; 2006). Lavender, geranium and clove essential oils do not protect *T. thermophila* cells against H2O2-induced oxidative stress. On the other hand, clove essential oil was significantly effective against nitrosative stress during the exponential phase. In the stationary phase, no protective effect was demonstrated against SNP-treated cells, although cypress, juniper and rosemary essential oils significantly increased cell numbers under oxidative stress.

According to our results, none of the 3 essential oils of lavender, geranium and thyme were able to reduce the two types of stress when used separately. We therefore wanted to assess the synergistic effect of the essential oils as an anti-stress agent. Very little information has been reported on the synergistic effect of essential oils as anti-stress agents. Thus, the mixture of 2 essential oils added to the culture medium of *T. thermophila* treated with H2O2 or SNP does not affect or protect the cells during the latent phase against oxidative and nitrosative stress. Although in the exponential growth phase, the 3 essential oils showed a highly significant increase in cell number compared with H2O2-treated cells, when mixed (Figure 24). On the other hand, these mixtures of essential oils (lavender-thyme, lavender-geranium and thyme-geranium) improved the growth of *T. thermophila* cells treated with SNP (Figure 25). The synergistic interaction does not appear to be selective for the type of stress and affects both oxidative and nitrosative stress. It seems that the anti-stress effect of the essential oils can be attributed to the synergistic interaction of the compounds making up the oils rather than to the use of the essential oil separately.

As lavender, geranium, thyme, rosemary, cypress, juniper and clove essential oils are used in traditional medicine and aromatherapy, our results, along with others, could be useful in considering the use of these essential oils as a functional food ingredient or dietary supplement for stress prevention in the future.

CONCLUSION

Stress in a дёпёгак way is consideredёrë as a risk factor for certain diseases of multiple origins. The link between stress and the onset of certain diseases is ëtabli via the body's inability to cope and put in place effective dëdures of dëfence when it is subjected to too much aggression. The frequency with which an organism is subjected to stressors, and their intensity, can lead the organism to exceed its adaptive capacities and thus put it in danger.

Oxidative stress is a complex and rapidly evolving concept. It is not a disease but a pathophysiological mechanism. The term oxidative stress indicates that the antioxidant status of cells is akre by exposure to oxidants. Today, it is recognised that oxidative stress is responsible for redox regulation involving reactive oxygen species (ROS) and reactive nitrogen species (RNS). Understanding the mëcanisms of action of ROS/RNS and their effects makes it possible to envisage prevention and/or anti-stress therapy. Thus, several studies using various cellular models have attempted to elucidate the involvement of ROS/RNS in the development of several diseases. Consequently, it has been formally proven that oxidative stress is implicated in the development of several human diseases, such as certain cardiovascular and neurological diseases, cancer, and diseases caused by inflammation and ageing. These studies, most of which are epidemiological, have shown that the use of antioxidants reduces the risk of certain diseases.

This is the context of our study, the origjnalk of which lies initially in the choice of the microorganism used as a cellular model, which is none other than *Tetrahymena thermophila*. This unicellular eukaryotic ^Hë protozoan has all the systems found in a human cell, such as the nucleus, mitochondria, cytoskeleton, sëcrëtion and chemical messenger response systems, which is why it was chosen as an alternative model to animal cells, particularly mammalian cells. The conservation of numerous intracellular mechanisms in eukaryotes makes it possible to study another cellular system in order to shed light on the biology of human cells. Thus, in our work we have used *T. thermophila* to ëvaluer the effects of oxidative stress and also the cellular responses to this stress which are universal in all eukaryotes. Although, each organism has its own '*style*' and its own '*secrets*' of adaptation. The second originalk of our work, which is both an appropriate and innovative idëe, is that of using natural essential oils to remëde the oxidative and nitrosative stress to which the protozoan used as a cellular model is exposed.

The results presented in this study support those reported in other research studies, which confirm the harmful effects of oxidative and nitrosative stress, particularly those caused by hydrogen peroxide and nitric oxide, on cells. These two stress agents induce major disturbances in cell growth, morphology and physiology. In order to better assess the metabolic disorders caused by these agents, we chose to assess their effects on an clë enzyme of carbohydrate metabolism, glyceraldehyde-3- phosphate dëshydrogënase (GAPDH). Thus and for the first time to our knowledge this enzyme has been purified and caractërisëed in *T. thermophila* to be used as a mëtabolic marker. Our aim was to study the effects of oxidative and nitrosative stress on this enzyme both *in vivo* and *in vitro*. Finally, in order to remëdiate the neiastic effects caused by this stress, we thoughtë of evaluating the antioxidant potential of certain essential oils in order to elucidate the effects of supplementation with these oils on the oxidative profile.

In practical terms, we can draw the following main conclusions from the results obtained:
- The sensitivity of *T. thermophila* to the 2 stressors (H2O2 and SNP) depends on the type

of stress and the concentration of the stressor. The dose-response curves differ from one stress agent to another.

- Morphological analysis of *T. thermophila* cells subjected to stress гёуёк an alteration in their shapes. This alteration is much more prononcëe in the case of SNP treatment than in that of H2O2 treatment.

- SNP induced a greater enzymatic antioxidant response than H2O2, in terms of both catalase and SOD activity. Similarly, lipid peroxidation was higher in the SNP treatment than in the H2O2 treatment.

- The specific activity of GAPDH, measured *in vivo*, decreased in both cases of stress. This decrease was more pronounced in the SNP treatment than in the H2O2 treatment. However, both stress agents induced overexpression of GAPDH.

- The purification of GAPDH from *T. thermophila* by fractional precipitation with ammonium sulphate followed by two column chromatographies (DEAE-cellulose and Mono-S) enabled us to characterise this enzyme in this protozoan for the first time.

- Biochemical characterisation of *T. thermophila* GAPDH revealed that it is a homotetrameric protein of 120 kDa, consisting of 4 identical subunits and presenting an isoenzymatic form with a *pI* of approximately 8.8.

- Determination of the physicochemical and kinetic parameters of purified GAPDH provided information on the optimal conditions for its enzymatic activity and its affinity for its two substrates.

- GAPDH activity is inhibited by both stressors. Comparison of IC50s determined after *in vitro* incubation of the purified enzyme with the stressor at different concentrations has rëvëlë a higher sensitivity of GAPDH a ГH2O2 than to SNP. Inhibition kinetics also differ between stressors.

- The two stress agents (H2O2 and SNP) also influence the growth of *T. thermophila* by reducing the number of cells during the latent and exponential phases. However, no significant influence was observed during the stationary phase.

- The antioxidant potential of essential oils is not apparent during the protozoa's first growth phase (latency phase). During this phase, *T. thermophila* cells adapt to the new culture conditions by synthesising the necessary enzymes.

- The protective effect of essential oils on the growth of *T. thermophila* exposed to oxidative and nitrosative stress only really becomes apparent from the exponential phase of the protozoan's growth. This effect seems to be influenced by the type of stress and the nature of the essential oil used.

- The effect of oxidative stress could be reduced during the exponential phase by essential oils of thyme, cypress, juniper and rosemary. The presence of these oils in the culture medium resulted in a sharp increase in the number of *T. thermophila* cells. In contrast, lavender, geranium and clove essential oils showed no oxidative anti-stress effect.

- The effect of nitrosative stress could also be attenuated in the presence of cypress, juniper, rosemary and clove essential oils during the exponential growth phase of *T. thermophila*. In contrast, lavender, thyme and geranium essential oils had no protective effect against nitrosative stress during this same growth phase.

- During the stationary phase, the different essential oils used did not show a protective effect against nitrosative stress. However, in the case of oxidative stress, the addition of cypress, juniper or rosemary essential oils improved cell numbers. It is potentially interesting to correlate the composition of essential oils with their anti-stress behaviour. These essential oils have different compositions, which makes it difficult to attribute the anti-stress effect to a specific compound.

- The use of lavender, thyme and geranium essential oils separately does not protect *T. thermophila* cells against the effects of oxidative stress and nitrosative stress.

- The blend of these essential oils (lavender, thyme and geranium) appears to have a protective effect against both types of stress during the exponential growth phase. This could be attributed to the synergistic interactions of the essential oils.

- The synergistic interaction does not appear to be selective with regard to the type of stress. It affects both oxidative and nitrosative stress.

While this ëstudy produced some interesting results, it opened up several avenues of research and left many questions unanswered. Our work suggests that there are many experimental prospects for answering these questions.

In general, this work will be followed up by an analysis of the fatty acid composition of *T. thermophila* membranes in response to oxidative and nitrosative stress, in order to determine the level of sensitivity of the cell.

It would be interesting to conduct a study on other enzymes in the glycolytic pathway in order to assess the impact of GAPDH inhibition *in vivo* on this pathway. In the same vein, the study of important enzymes in cellular metabolism will help to elucidate the physiological consequences caused by stress and thus identify stress proteins in order to understand the defence mechanism used by *T. thermophila*.

Another possible approach would also be interesting, and that would be to carry out a study at the molecular level to obtain information on the gënes of resistance to oxidative and nitrosative stress. Similarly, the development of the tertiary structure of GAPDH would be desirable in order to understand the interactions between the stress agent and the enzyme.

On the other hand and to better assess the anti-stress power of essential oils, similar more promising studies targeting other biological markers, such as the evaluation of lipid peroxidation, antioxidant enzymes and other metabolic enzymes will be undertaken on our cellular modële and/or another modële.

Finally, we are also looking at the chemical characterisation of essential oils used as pharmaceutical products, with a view to carrying out subsequent studies on the use of isolated essential oil compounds to target compounds with oxidative or nitrosative anti-stress properties. In-depth studies on the pharmacokinetics and pharmacodynamics of the active ingredients would also be useful for determining preventive and therapeutic doses.

BIBLIOGRAPHICAL REFERENCES

Acharya, A., Das, I., Chandhok, D., Saha, T. (2010). Redox regulation in cancer: a doubledged sword with therapeutic potencial. *Oxid. Med. Cell. Longev.* 3, 23-34.

Aebi, H. (1984). Catalase *in vivo. Methods Enzymol.* 105, 121-126.

Ahmad, Q.R., Nguyen, D.H., Wingerd, M.A., Church, G.M., Steffen, M.A. (2005). Molecular weight assessment of proteins in total proteome profiles using ID-PAGE and LC/MS/MS. *Proteome Science*, 3, 1-7.

Allen, K.G., Banthorpe, D.V., Charlwood, B.V., Voller, C.M. (1977). Biosynthesis of Artemisia ketone in higher plants. *Phystochem.* 16, 79-83.

Allen, R.D. (1967). Fine structure, reconstruction, and possible functions of components of the cortex of *Tetrahymena pyriformis. J. Protozool.* 14, 553-565.

Ames, B. N., Shigenaga, M. K., Hagen, T. M. (1993). Oxidants, antioxidants, and the degenerative diseases of aging. *Proc. Natl. Acad. Sci.* USA. 90 (17), 7915-7922.

Anand, P., Stamler, J.S. (2012). Enzymatic mechanisms regulating protein S-nitrosylation: implication in health and disease. *J. Mol. Med.* 90(3), 233-244.

Atanda, O.O., Akpan, I., Oluwafemi, F. (2007). The potential of some spice essential oils in the control of A. Parasiticus CFR 223 and aflatoxin production. *Food Control.* 18, 601-607.

Aufderheide, K.J. (1979). Mitochondrial associations with specific microtubular components of the cortex of *Tetrahymena thermophila.* I. Cortical patterning of mitochondria. *J. Cell. Sci.* 39, 299-312.

Baek, D., Jin, Y., Jeong, J.C., Lee, H.Y., Moon, H., Lee, J., Shin, D. (2008). Suppression of reactive oxygen species by glyceraldehyde-3-phosphate dehydrogenase. *Phytochem.* 69, 333338.

Baibai, T., Oukhattar, L., Moutaouakkil, A., Soukri, A. (2007). Purification and characterization of glyceraldehyde-3-phosphate dehydrogenase from European pilchard *Sardina pilchardus. Acta Biochim. Biophys. Sin.* 39, 947-954.

Bakkali, F., Averbeck, S., Averbeck, D., Idaomar, M. (2007). Biological effects of essential oils- A review. *Food chem. toxicol.* 46, 446-475.

Bakkali, F., Averbeck, S., Averbeck, D., Zhiri, A., Baudoux, D., Idaomar, M. (2006). Antigenotoxic effects of three essential oils in diploid yeast (*Saccharomyces cerevisiae*) after treatments with UVC radiation, 8- MOP plus UVA and MMS. *Mutat. Res.* 606, 27-38.

Bakkali, F., Averbeck, S., Averbeck, D., Zhiri, A., Idaomar, M. (2005). Cytotoxicity and gene induction by some essential oils in the yeast *Saccharomyces cerevisiae.* Mutat. Res. 585, 113.

Beckman, J.S. (1996). Oxidative damage and tyrosine nitration from peroxynitrite. *Chem. Res. Toxicol.* 9, 836-844.

Beckman, J.S., Beckman, T.W., Chen, J., Marshall, P.A., Freeman, B.A. (1990). Apparent hydroxyl radical production by peroxynitrite: implications for endothelial injury from nitric oxide and superoxide. *Proc. Natl. Acad. Sci.* 87, 1620-1624.

Beckman, K.B., Ames, B.N. (1998). The free radical theory of aging matures. *Physiol. Rev.* 78, 547-581.

Ben Saida, F.W. (2007). Les huiles essentielles et leur utilisation en usage tliératique. Master memory, Institut Supërieur de Biotechnologie, Universite de Monastir.

Berger, M.M. (2006). Nutritional manipulation of oxidative stress: state of the art. *Nutr. Clin. Metabol.* 20, 48-53.

Berlette, B.S., Stadtman, E.R. (1997). Protein oxidation in aging disease, and oxidative stress.

J. Biol. Chem. 272, 20313-20316.

Besson-Bard, A., Pugin, A., Wendehenne, D. (2008). New insights into oxide signaling in plants. *Annu. Rev. Plant. Biol.* 59, 21-39.

Bielski, B.H.J., Arudi, R.L., Sutherland, M.W. (1983). A study of the activity of HO_2^-/O_2^- with unsaturated fatty acids. *J. Biol. Chem.* 258, 4759-4761.

Biriukova, E. N., Medentsev, A.G., Arinbasarova, A. Yu., Akimenko, V.K. (2006). Tolerance of the yeast *Yarrowia lipolytica* to oxidative stress. *Microbiol.* 75, 293-298.

Bliss, C. I. (1935). The calculation of the dosage mortality curve. *Ann. Appl. Biol.* 22, 134165.

Bonnefont-Rousselot, D., Therond, P., Delattre, J. (2003). Free radicals and antioxidants. In: Biochimie pathologique : aspects moleculaires et cellulaires. Delattre, J., Durant, G., Jardillier, J.C. Eds: Mcdecine-sciences. Flammarion (Paris), Pp 59-81.

Boveris, A., Cadenas, E. (1975). Mitochondrial production of superoxide anions and its relationship to the antimycin-insensitive respiration. *FEBS Lett.* 54, 311-314.

Boveris, A., Chance, B. (1973). The mitochondrial generation of hydrogen peroxide. General properties and effect of hyperbaric oxygen. *Biochem. J.* 134, 707-716.

Boyd, D.A., Cvitkovitch, D.G., Hamilton, I.R. (1995). Sequence, expression, and function of the gene for the non-phosphorylating, NADP-dependent glyceraldehyde-3-phosphate dehydrogenase of *Streptococcus mutans*. *J. Bacteriol.* 177, 2622-2627.

Bradford, M. (1976). A rapid and sensitive method for the quantitation of microgram quantities of protein utilizing the principle of protein dye binding. *Anal. Biochem.* 72, 248254.

Branny, P., De la Torre, F., Garel, J.R. (1998). An operon encoding three glycolytic enzymes in *Lactobacillus delbrueckii subsp. bulgaricus*: glyceraldehyde-3-phosphate dehydrogenase, phosphoglycerate kinase and triosephosphate isomerase. *Microbiol.* 144, 905-914.

Breen, A.P., Murphy, J.A. (1995). Reactions of oxyl radicals with DNA. Free *Rad. Biol*. *Med.* 18 (6), 1033-1077.

Brinkmann, H., Cerff, R., Salomon, M., Soll, J. (1989). Cloning and sequence analysis of cDNAs encoding the cytosolic precursors of subunits *GapA* and *GapB* of chloroplast glyceraldehyde-3-phosphate dehydrogenase from pea and spinach. *Plant. Mol. Biol.* 13, 8194.

Briviba, K., Klotz, L.O., Sies, H. (1997). Toxic and signalling effects of photochemically or chemically generated singlet oxygen in biological systems. *J. Biol. Chem.* 378, 1259-1265.

Bruneton, J. (1999). Pharmacognosy. Phytochemistry of medicinal plants. Technique et Documentation Lavoisier. 2nd edition. Paris. 915.

Buchbauer, G., Jirovetz, L. (1994). Aromatherapy-Use of fragrances and essential oils as medicaments. *Flav. Frag. J.* 9, 217-222.

Bush, P. A., Gonzalez, N. E., Griscavage, J. M., Ignarro, L. J. (1992). Nitric oxide synthase from cerebellum catalyzes the formation of equimolar quantities of nitric oxide and citrulline from L-arginine. *Biochem. Biophys. Res. Comm.* 185 (3), 960-966.

Cabiscol, E., Piulats, E., Echaves, P., Herrero, E., Ros, J. (2000). Oxidative stress promotes specific protein damage in *Saccharomyces cerevisiae*. *J. Biol. Chem.* 275 (35), 27393-27398.

Cannon, W.B. (1932). The Wisdom of the Body. W.W. Norton & company, New York.

Cerff, R. (1982). Separation and purification of NAD- and NADP-linked glyceraldehyde-3-phosphate dehydrogenase from higher plants. In: Methods in chloroplast molecular biology, pp 683-694. Edelman, M., Hallick, R.B. y Chua, N.-H. (ed) Elsevier Biomedical Press, Amsterdam.

Charpentier, B., Bardey, V., Robas, N., Banlant, C. (1998). The EIIGlc protein is involved in glucose-mediated activation of *Escherichia coli* gaper and gapB-pgk transcription. *J.*

Bacteriol. 180, 6476-6483.

Cheeseman, K. H., Slater, T. F. (1993). An introduction to free radical biochemistry. *Br. Med. Bull.* 49 (3), 588-603.

Circu, M.L., Aw, T.Y. (2010). Reactive oxygen species, cellular redox system, and apoptosis. *Free. Radic. Biol. Med.* 48, 749-762.

Clergeaud, C.L. (2000). Les huiles végétales, huiles de santé et de beaute. Ed. Atlantica, Paris. 44-45.

Cohen, G., Hochstein, P. (1963). Glutathione peroxidase: the primary agent for the elimination of hydrogen peroxide in erythrocytes. *Biochem.* 2, 1420-1428.

Colussi, C., Albertini, M.C., Coppola, S., Rovidati, S., Galli, F., Ghibelli, L. (2000). H2O2-induced block of glycolysis as an active ADP-ribosylation reaction protecting cells from apoptosis. *FASEB J.* 14, 2266-2276.

Corliss, J.O. (1952). Systematic status of the pure culture ciliate known as "*Tetrahymena geleii*" and "Glaucoma piriformis". *Sci.* 116, 188-191.

Crane, B.R., Sudhamsu, J., Patel, B.A. (2010). Bacterial nitric oxide synthases. *Annu. Rev. Biochem.* 79, 445-470.

Cross, A. R., Jones, O. T. (1991). Enzymic mechanisms of superoxide production. *Biochim Biophys Acta.* 1057 (3), 281-298.

Davies, K.J. (2000). Oxidative stress, antioxidant defences, and damage removal, repair, and replacement systems. *IUBMB Life.* 50, 279-289.

Deans, S.G., Ritchie, G. (1987). Antibacterial properties of plant essential oils. *Int. J. Food Microbiol.* 5, 165-180.

DeDuve, C., Baudhuin, P. (1966). Peroxisomes (microbodies and related particles). *Physiol Rev.* 46, 323-357.

Delgado, M.L., O'Connor, J.E., Azorin, I., Renau-Piqueras, J., Gil, M.L., Gozalbo, D. (2001). The glyceraldehyde-3-phosphate dehydrogenase polypeptides encoded by the *Saccharomyces cerevisiae* TDH1, TDH2 and TDH3 genes are also cell wall proteins. *Microbiol.* 147, 411417.

Densiov, E.T., Afanas'ev, I.B. (2005). In: Oxidation and antioxidants in organic chemistry and biology. Eds: Taylor & Francis group (U.S.A), Pp 703-861.

Dias, J., Sanchez, M.J., Navarro, A. (1998). Peroxidacion lipidica en neonatologia. *Pediatrika.* 8 (6), 221-232.

Dias, N., Mortara, R. A., Lima, N. (2003). Morphological and physiological changes in *Tetrahymena pyriformis* for the *in vitro* cytotoxicity assessment of Triton X-100. *Toxicol. In vitro.* 17, 357-366.

Dimmeler, S., Lottspeich, F., Brune, B. (1992). Nitric oxide causes ADP-ribosylation and inhibition of glyceraldehyde-3-phosphate dehydrogenase. *J. Biol. Chem.* 267 (24), 1677116774.

Dowling, D.K., Simmons, L.W. (2009). Reactive oxygen species as universal constraints in life-history evolution. *Proc. Biol. Sci.* 276, 1737-1745.

Edreva, A. (2005). Generation and scavenging of reactive oxygen species in chloroplasts: a submolecular approach. Agric. *Ecosyst. Environ.* 106, 119-133.

Eikmanns, B.J. (1992). Identification, sequence analysis, and expression of a *Corynebacterium glutamicum* gene cluster encoding the three glycolytic enzymes glyceraldehyde-3-phosphate dehydrogenase, 3-phosphoglycerate kinase, and triosephosphate isomerase. *J. Bacteriol.* 174, 6076-6086.

Elliott, A.M., Gruchy, D.F. (1952). The occurrence of mating types in *Tetrahymena*. Biol.

Bull (Woods, Hole, Mass). 103, 301.

Errafiy, N., Soukri, A. (2012). Purification and partial characterization of glyceraldehyde-3-phosphate dehydrogenase from the ciliate *Tetrahymena thermophila*. *Acta. Biochim. Biophys. Sin.* 44, 527-534.

Evans, W.J. (2000). Vitamin E, vitamin C, and exercise. *Am. J. Clin. Nutr.* 72, 647S-652S.

Faleiro, L., Miguel, G.M., Guerrero, C.A.C., Brito, J.M.C. (1999). Antimicrobial activity of essential oils of *Rosmarinus officinalis* L., *Thymus mastichina* (L) L. ssp. mastichina and *Thymus albicans*. In: Proceedings of the II WOCMAP congress on medicinal and aromatic plants, part 2: pharmacognosy, pharmacology, phytomedicine, toxicology.

Faure-Fremiet, E., Gauchery, M., Rouiller, C. (1956). Ciliary origin of pedunculated scleroproteic fibers in ciliated peritricha; electon microscopic study. *Exp. Cell. Res.* 11, 527541.

Favier, A. (2003). Oxidative stress. Interet conceptuel et expérimental dans la comprehension des mecanismes des maladies et potentiel therapeutique. *Actual chim.* N° 269270, 108-115.

Fermani, S., Sparla, F., Falini, G., Martelli, P.L., Casadio, R., Pupillo, P., Ripamonti, A., Trost, P. (2007). Molecular mechanism of thioredoxin regulation in photosynthetic A2B2-glyceraldehyde-3-phosphate dehydrogenase. *Proc. Natl. Acad. Sci.* 104 (26), 11109-11114.

Figge, R.M., Schubert, M., Brinkman, H., Cerff, R. (1999). Glyceraldehyde-3-phosphate dehydrogenase gene diversity in eubacteria an eukaryotes: Evidence for intra- and interkingdom gene transfer. *Mol. Biol. Evol.* 16, 440-429.

Fillinger, S., Boschi-Muller, S., Azza, S., Dervyn, E., Branlant, G. (2000). Two glyceraldehydes-3-phosphate dehydrogenases with opposite physiological roles in a nonphotosynthetic bacterium. *J. Biol. Chem.* 275, 14031-14037.

Finaud, J., Lac, G., Filaire, E. (2006). Oxidative stress: relationship with exercise and training. *Sports Med.* 36, 327-358.

Florini, J.R., Vestling, C.S. (1957). Graphical determination of the dissociation constants for two-substrate enzyme systems. *Biochim. Biophys. Acta.* 25, 575-578.

Foote, Ch.S. (1982). Light, oxygen and toxicity. In: Pathology of oxygen. Anne, P (eds). Academia Press. London. 21-44.

Forthergill-Gilmore, L.A., Michels, P.A.M. (1993). Evolution of glycolysis. *Prog. Biophys. Mol. Biol.* 59, 105-235.

Foster, M.W., Hess, D.T., Stamler, J.S. (2009). Protein S-nitrosylation in health and disease: a current perspective. *Trends Mol. Med.* 15 (9), 391-404.

Fourrat, L., Iddar, A., Soukri, A. (2007a). Purification and characterization of a cytosolic Glyceraldehyde-3-phosphate dehydrogenase from the dromedary camel. Acta Biochim, Biophys, Sin, 39 (2), 148-154.

Fourrat, L., Iddar, A., Valverde, F., Serrano, A., Soukri, A. (2007b). Effects of oxidative and nitrosative stress on *Tetrahymena pyriformis* glyceraldehyde-3-phosphate dehydrogenase. *J. Eukaryot. Microbiol.* 54 (4), 338-346.

Franchomme, P., Pёnoёl, D. (1990). I .'aromatherapie exactement. Encyclopёdie de l'utilisation therapeutique des huiles essentielles. Roger Jallois editeur. Limoges. 445.

Franke, W.W., Eckert, W.A., Krien, S. (1971). Cytomembrane differentiation in a ciliate, *Tetrahymena pyriformis*. I. Endoplasmic reticulum and dictyosome equivalents. *Z. Zellforsch. Mikrosk. Anat.* 119, 577-604.

Frei, B. (1994). Reactive oxygen species and antioxidant vitamins: Mechanisms of action. *Am. J. Med.* 97 (3A), 5S-13S.

Fridovich, I. (1986). Oxygen radicals, hydrogen peroxide and oxygen toxicity. Free radicals and biology. Pryor, W. A. New York, Academic Press. 1, 239-246.

Fridovich, I. (1997). Superoxide anion radical, superoxide dismutases, and related matters. *J. Biol. Chem.* 272 (30), 18515-18517.

Furgason, W.H. (1940). The significant cytostomal pattern of the "Glaucoma-Colpidium group" and a proposed new genus and species, *Tetrahymena gelei. Arch. Protistenkd.* 94, 224266.

Gavin, R.H. (1965). The effects of heat and cold on cellular development in *Tetrahymena pyriformis* WH-6. *J. Protozool.* 12, 307-318.

Gerber, M., Boutron-Ruault, M. C., Hercberg, S., Riboli, E., Scalbert, A., Siess, M. H. (2002). Food and cancer: state of the art about the protective effect of fruits and vegetables. Bull. *Cancer.* 89, 293-312.

Giugliano, D., Ceriello, A., Paolisso, G. (1996). Oxidative stress and diabetic vascular complications. *Diabetes Care* 19 (3), 257-267.

Glaser, P.E., Gross, R.W. (1995). Rapid plasmenylethanolamine-selective fusion of membrane bilayers catalyzed by an isoform of glyceraldehyde-3-phosphate dehydrogenase: discrimination between glycolytic and fusogenic roles of individual isoforms. *Biochem.* 34 (38), 12193-12203.

Grant, C.M., Quinn, K.A., Dawes, I.W. (1999). Differential protein S-thiolation of glyceraldehyde-3-phosphate dehydrogenase isoenzymes influences sensitivity to oxidative stress. *Mol. Cell. Biol.* 19 (4), 2650-2656.

Griendling, K.K., Sorescu, D., Ushio-Fukai, M. (2000). NAD(P)H oxidase: role in cardiovascular biology and disease. *Circ Res.* 86 (5), 494-501.

Grisson, F.C., Kahn, J.S. (1975). Glyceraldehyde-3-phosphate dehydrogenase from *Euglena gracilis*. Purification and physical and chemical characterization. *Arch. Biochem. Biophys.* 171, 444-458.

Guenther, E. (1972). The Essential Oils. Robert E. Krieger Publishing Company; Inc, Huntington, New York. vol l.

Gutteridge, J.M., Halliwell, B. (2000). Free radicals and antioxidants in the year 2000. A historical look to the future. *Ann. NY Acad. Sci.* 899, 136-147.

Habenicht, A. (1997). The non-phosphorylating glyceraldehyde-3-phosphate dehydrogenase: biochemistry, structure, occurrence and evolution. *Biol. Chem.* 378, 1413-1419.

Hafid, N., Valverde, F., Villalobo, E., Elkebbaj, M.S., Torres, A., Soukri, A., Serrano, A. (1998). Glyceraldehyde-3-phosphate dehydrogenase from *Tetrahymena pyriformis*: enzyme purification and characterization of a *gap C* gene with primitive eukaryotic features. *Comp. Biochem. and Physiol. B.* 119, 493-503.

Halliwell, B. (1992). Reactive oxygen species and the central nervous system. *J. Neurochem.* 59 (5), 1609-1623.

Halliwell, B. (1996). Antioxidants in human health and disease. *Annu. Rev. Nutr.* 16, 33-50.

Halliwell, B., Chirico, S. (1993). Lipid peroxidation: its mechanism, measurement and significance. *Am. J. Clin. Nutr.* 57, 715-722.

Halliwell, B., Gutteridge, J.M.C. (1999). Free radicals in biology, and medicine. 3ed. Oxford, New York: Oxford Science Publications. Oxford University Press. p. 936.

Halliwell, B., Whiteman, M. (2004). Measuring reactive species and oxidative damage in vivo and in cell culture: how should you do it and what do the results mean? *Br. J. Pharmacol.* 142, 31-32.

Hanafy, K.A., Krumenacker, J.S., Murad, F. (2001). NO, nitrotyrosine, and cyclic GMP in Signal transduction. *Med. Sci. Monit*. 7, 801-819.

Hara, M.R., Agrawal, N., Kim, S.F., Cascio, M.B., Fujimuro, M., Ozeki, Y., Takahashi, M., Cheah, J.H., Tankou, S.K., Hester, L.D., Hayward, S.D., Snyder, S.H., Sawa, A. (2005). S-nitrosylated GAPDH initiates apoptotic cell death by nuclear translocation following Siah1 binding. *Nat. Cell. Biol*. 7 (7), 665-674.

Harman, D. (1956). Aging: a theory based on free radical and radiation chemistry. *J. Gerontol*. 11, 298-300.

Harman, D. (1981). The aging process. *Proc. Narl. Acad. Sci*. USA. 78, 7124-7128.

Harris, J.I., Waters, M. (1976). The enzymes, 3rd Ed., edited by P. D. Boyer, ch. 13. New York: Academic press.

Hess, D.T., Matsumoto, A., Kim, S.O., Marshall, H.E., Stamler, J.S. (2005). Protein S-nitrosylation: Purview and parameters. *Nat. Rev. Mol. Cell. Biol*. 6, 150-166.

Hoffmann, J., Dimmeler, S., Haendeler, J. (2003). Shear stress increases the amount of S-nitrosylated molecules in endothelial cells:important role for signal transduction. *FEBS Lett*. 551, 153-158.

Holz, G.G.Jr. (1973). The nutrition of *Tetrahymena thermophila*: Essential nutrients, feeding, and digestion. In: Biology of Tetrahymena (A.M. Elliott, ed.). Dowden, Hutchinson & Ross. Stroudsburg. Pa. Pp. 89-98.

Hood, W., Carr, N.G. (1969). Association of NAD and NADP linked glyceraldehyde-3-phosphate dehydrogenase in the blue-green alga, *Anabaena variabilis*. *Planta*. 86, 250-258.

Iglesias, A., Losada, M. (1988). Purification and kinetic and structural properties of spinach leaf NADP-dependent nonphosphorylating glyceraldehyde-3-phosphate dehydrogenase. *Arch. Biochem. Biophys*. 260, 830-840.

Iglesias, A., Serrano, A., Guerrero, M.G., Losada, M. (1987). Purification and properties of NADP-dependent non-phosphorylating glyceraldehyde-3-phosphate dehydrogenase from the green alga *Chlamydomonas reinhardtii*. *Biochim. Biophys. Acta* 925, 1-10.

Inal, M.E., Kanbak, G., Sunal, E. (2001). Antioxidant enzyme activities and malondialdehyde levels related to aging. *Clin. Chim. Acta*. 305 (1-2), 75-80.

Ishii, T., Sunami, O., Nakajima, H., Nishio, H., Takeuchi, T., Hata, F. (1999). Critical role of sulfenic acid formation of thiols in the inactivation of glyceraldehyde-3-phosphate dehydrogenase by nitric oxide. *Biochem. Pharmacol*. 58 (1), 133-43.

Jaeckel-Williams, R. (1978). Nuclear divisions with reduced numbers of microtubules in *Tetrahymena*. *J. Cell. Sci*. 34, 303-319.

Ji, L.L., Mitchell, E.W. (1992). Glutathione and antioxidant enzymes in skeletal muscle: effects of fiber type and exercise intensity. *J. Appl. Physiol*. 73, 1854-1859.

Jung, T., Bader, N., Grune, T. (2007). Oxidized proteins: intracellular distribution and recognition by the proteasome. *Arch. Biochem. Biophys*. 462, 231-237.

Kaczanowska, J., Buzanska, L., Ostrowski, M. (1993). Relationship between spatial patterning of basal bodies and membrane skeleton (epiplasm) during the cell cycle of *Tetrahymena*: cdaA mutant and anti-membrane immunostaining. *J. Eukryot. Microbiol*. 40, 747-754.

Kaneko, T., Sato, S., Kotani, H., Tanaka, A., Asamizu, E., Nakamura, Y., Miyajima, N., Hirosawa, M., Sugiura, M., Sasamoto, S., Kimura, T., Hosouchi, T., Matsuno, A., Muraki, A., Nakazaki, N., Naruo, K., Okunura, S., Shimpo, S., Takeuchi, C., Wada, T., Watanabe, A., Yamada, M., Yasuda, M. and Tabata, S. (1996). Sequence analysis of the genome of the unicellular cyanobacterium *Synechocystis* sp. strain PCC 6803. II. Sequence determination of the entire genome and assignment of potential protein-coding regions. *DNA Res.* 3, 109-136.

Kiy, T., Tiedtke, A. (1992). Mass cultivation of *Tetrahymena thermophila* yielding high cell densities and short generation times. *Appl. Microbiol. Biotechnol.* 37, 576-579.

Koch, C., Reichling, J., Schneele, J., Schnitzler, P. (2008). Inhibitory effect of essential oils against herpes simplex virus type 2. *Phytomedicine.* 15, 71-78.

Koechlin-ramonatxo, C. (2006). Oxygen, oxidative stress and antioxidant supplementation, or another way for nutrition in respiratory diseases. *Nutr. Clin. Metabol.* 20, 165-177.

Kohen, R., Nyska, A. (2002). Oxidation of biological systems: oxidative stress phenomena, antioxidants, redox reactions and methods for their quantification. *Toxicol. Pathol.* 30, 620650.

Komanec, J., Lempel'ova, A., Novakova, R., Rezuchova, B., Homerova, D. (1997). Expression of the *Streptomyces aureofaciens* glyceraldehyde-3-phosphate dehydrogenase gene (gap) is developmentally regulated and induced by glucose. *Microbiol.* 143, 3555-3561.

Korycka-Dahi, M., Richardson, T. (1981). Initiation of oxidative changes in food. Symposium: oxidative changes in milk. *J. Diary Sci.* 63 (7), 1181-1208.

Kosar, M., Ozek, T., Coger, F., Kurkcuoglu, M., Husnu Can, K. (2005). Comparison of microwave-assisted hydrodistillation and hydrodystillation methods for the analysis of volatile secondary metabolites. *Pharm. Biol.* 6, 491-495.

Kots, A.Y., Skurat, A.V., Sergienko, E.A., Bulargina, T.V., Severin, E.S. (1992). Nitroprusside stimulates the cysteine-specific mono (ADP-ribosylation) of glyceraldehyde-3-phosphate dehydrogenase from human erythrocytes. *FEBS Lett.* 300 (1), 9-12.

Kunst, F., Ogasawara, N. Moszer, I. et *al.* (1997). The complete genome sequence of the gram-positive bacterium *Bacillus subtilis*. *Nature.* 390, 249-256.

Kurz, S., Tiedtke, A. (1993). The Golgi apparatus of *Tetrahymena thermophila*. *J. Eukaryot. Microbiol.* 40, 10-13.

Laemmli UK (1970). Cleavage of structural proteins during assembly of the head of bacteriophage T4. *Nature.* 227, 680-685.

Lansing, T.J., Frankel, J., Jenkins, L.M. (1985). Oral ultrastructure and oral development in the misaligned undulating membrane mutant of *Tetrahymena thermophila*. *J. Protozool.* 31, 126-139.

Lee, H.C., Yin, P.H., Lu, C.Y., Chi, C.W., Wei, Y.H. (2000) Increase of mitochondria and mitochondrial DNA in response to oxidative stress in human cells. *Biochem. J.* 348, 425-432.

Lehucher-Michel, M.P., Lesgards, J.F., Delubac, O., Stocker, P., Durand, P., Prost, M. (2001). Stress oxydant et pathologies humaines. *Lapresse medicale.* 30, 1076-1081.

Li, A.D., Anderson, L.E. (1997). Expression and characterization of pea chloroplastic glyceraldehyde-3-phosphate dehydrogenase composed of only the B-subunit. *Plant Physiol.* 115, 1201-1209.

Li, A.D., Stevens, F.J., Huppe, H.C., Kersanach, R., Anderson, L.E. (1997). *Chlamydomonas reinhardtii* NADP-linked glyceraldehydes-3-phosphate dehydrogenase contains the cysteine residues identified as potentially domain-locking in the higher plant enzyme and is light activated. *Photosynth. Res.* 51, 167-177.

Li, W.G., Miller, F.J.Jr., Zhang, H.J., Spitz, D.R., Oberley, L.W., Weintraub, N.L. (2001). H_2O_2-induced O_2^{-} production by a non-phagocytic NAD(P)H oxidase causes oxidant injury. *J. Biol. Chem.* 276 (31), 29251-29256.

Lineweaver M and Burk D. (1934). The determination of enzyme dissociation constants. *J.*

Am. Chem. Soc. 56, 658-666.

Loomis, D., and Croteau, R. (1980). *Biochemistry* of Terpenoids: A Comprehensive Treatise. In: P. K. Stumpf and E. E. Conn (eds.) the Biochemistry of Plants. Lipids: Structure and Function. Academic Press, San Francisco. 4, 364-410.

Lucchesi, M.E., Smadja, J., Bradshaw, S., Louw, W., Chemat, F. (2007). Solvent free microwave extraction of *Elletaria cardamomum* L: A multivariate study of a new technique for the extraction of essential oil. *J. Food Engineer.* 79,1079-1086.

Machado, M., Dinis, A.M., Salgueiro, L., Custodio, Josë B.A., Cavaleiro, C., Sousa, M.C. (2011). Anti-Giardia activity of *Syzygium aromaticum* essential oil and eugenol: Effects on growth, viability, adherence and ultrastructure. *Exp. Parasitol.* 127 732-739.

Maciel, M.V., Morais, S.M., Bevilaqua, C.M.L., Silva, R.A., Barros, R.S., Sousa, R.N., Sousa, L.C., Brito, E.S., Souza-Neto, M.A. (2010). Chemical composition of Eucalyptus spp. essential oils and their insecticidal effects on Lutzomyia longipalpis. *Vet. Parasitol.* 167, 1-7.

Martin, W., Brinkmann, H., Savona, C., Cerff, R. (1993). Evidence for a chimeric nature of nuclear genomes: Eubacterial origin of eukaryotic glyceraldehyde-3-phosphate dehydrogenase genes. *Proc. Natl. Acad. Sci.* U.S.A. 90, 8692-8696.

Martin, W., Cerff, R. (1986). Prokaryotic features of a nucleus-encoded enzyme. cDNA sequences for chloroplast and cytosolic glyceraldehyde-3-phosphate dehydrogenases from mustard (*Sinapis alba*). *Eur. J. Biochem.* 159, 323-331.

Martinez-Cayuela, M. (1995). Oxygen free radicals and human disease. *Biochem.* 77, 147161.

McDonald, B.B. (1962). Synthesis of deoxyribonucleic acid by micro- and macronuclei of *Tetrahymena pyriformis. J. Cell. Biol.* 13, 193-203.

Meilhoc, E., Cam, Y., Skapski, A., Bruand, C. (2010). The response to nitric oxide of the nitrogen-fixing symbiont Sinorhizobium meliloti. *Mol. Plant. Microbe Interact.* 23, 748-759.

Meyer, R.R., Boyd, C.R., Rein, D.C., Keller, S.J. (1972). Effects of ethidium bromide on growth and morphology of Tetrahymena pyriformis. *Exp. Cell. Res.* 70 (1), 233-237.

Meyer-Gauen, G., Herbrand, H., Cerff, R., Martin, W. (1998). Gene structure, expression in *Escherichia coli* and biochemical properties of the NAD^+ -dependent glyceraldehyde-3-phosphate dehydrogenase from *Pinus silvestris* chloroplasts. *Gene.* 209, 167-174.

Meyer-Gauen, G., Schnarrenberger, C., Cerff, R., Martin, W. (1994). Molecular characterization of a novel, nuclear-encoded, NAD^+ -dependent glyceraldehyde-3-phosphate dehydrogenase in plastids of the gymnosperm *Pinus sylvestris* L. *Plant. Mol. Biol.* 26, 11551166.

Meyer-Siegler, K., Mauro, D.J., Seal, G., Wurzer, J., deRiel, J.K., Sirover, M.A. (1991). A human nuclear uracil DNA glycosylase is the 37-kDa subunit of glyceraldehyde-3-phosphate dehydrogenase. *Proc. Natl. Acad. Sci.* USA. 88 (19), 8460-8464.

Mishra P. K., Shukla R., Singh P., Prakash B., Dubey N. K. (2012). Antifungal and antiaflatoxigenic efficacy of Caesulia axillaris Roxb. essential oil against fungi deteriorating some herbal raw materials, and its antioxidant activity. *Ind. Crop. Prod.* 36, 74-80.

Mohr, S., Hallak, H., De Boitte, A., Lapetina, E.G., Brune, B. (1999). Nitric Oxide-induced S-Glutathionylation and Inactivation of Glyceraldehyde-3-phosphate Dehydrogenase. *J. Biol. Chem.* 274 (14), 9427-9430.

Mojica, E.R., Deocaris, C.C., Endriga, M.A. (2004). Essential oils as anti-protozoal agents. *Philipp. J. Crop. Sci.* 29 (3), 41-43.

Moncada, S., Palmer, R. M., Higgs, E. A. (1991). Nitric oxide: physiology, pathophysiology and pharmacology. *Pharmacol. Rev.* 43, 109-142.

Morigasaki, S., Shimada, K., Ikner, A., Yanagida, M., Shiozaki, K. (2008). Glycolytic enzyme GAPDH promotes peroxide stress signalling through multistep phosphorelay to a MAPK cascade. *Mol. Cell.* 30 (1), 108-113.

Mounaji, K., Erraiss, N.E., Iddar, A., Wegnez, M., Serrano, A., Soukri, A. (2002). Glyceraldehyde-3-phosphate dehydrogenase from the newt *Pleurodeles waltl.* Protein purification and characterization of a *gap C* gene. *Comp. Biochem. Physiol. B.* 131, 411-421.

Mountassif, D., Baibai, T., Fourrat, L., Moutaouakkil, A., Iddar, A., ElKebbaj, M.S., Soukri, A. (2009). Immunoaffinity purification and characterization of glyceraldehyde-3-phosphate dehydrogenase from human erythrocytes. *Acta Biochim. Biophys. Sin.* 44 (5), 399-406.

Mountassif, D., Kabine, M., Manar, R., Bourhim, N., Zaroual, Z., Latruff, N., El Kebbaj, M. (2007). Physiological, morphological and metabolic changes in *Tetrahymena pyriformis* for the *in vivo* cytotoxicity assessment of metallic pollution: Impact on D-B-hydroxybutyrate dehydrogenase. *Ecol. Indic.* 7, 882-894.

Mourey, A., Canillac, N. (2002). Anti-Listeria monocytogenes activity of essential oils components of conifers. *Food Control.* 13, 289- 292.

Muzykantov, V.R. (2001). Targeting of superoxide dismutase and catalase to vascular endothelium. *J. Control. Release.* 71 (1), 1-21.

Nanney, D.L., McCoy, J.W. (1976). Characterization of the species of the *Tetrahymena pyriformis* complex. *Trans. Am. Microsc. Soc.* 95, 664-682.

Nilsson, J.R. (1979). Phagotrophy in Tetrahymena. In "Biochemistry and Physiology of Protozoa" (Levandowsky, M and Hunter, S.H., eds), 2nd ed, Academic Press, New York. Vol. 2, pp. 339-379.

Packer, L., Tritschler, H.J., Wessel, K. (1997). Neuroprotection by the metabolic antioxidant alpha-lipoic acid. *Free Radic. Biol. Med.* 22, 359-378.

Paoletti, F., Aldinucci, D., Mocali, A., Carparrini, A. (1986). A sensitive spectrophotometric method for the determination of superoxide dismutase in tissue extracts. *Anal. Biochem.* 154 (2), 526-541.

Patterson, R.L., van Rossum, D.B., Kaplin, A.L., Barrow, R.K., Snyder, S.H. (2005). Inositol 1,4,5-trisphosphate receptor/GAPDH complex increases Ca^{2+} release via locally derived NADH. *Proc. Natl. Acad. Sci.* USA. 102 (5), 1357-1359.

Perry, N.S., Bollen, C., Perry, E.K., Ballard, C. (2003). Salvia for dementia therapy: review of pharmacological activity and pilot tolerability clinical trial. *Pharmacol. Biochem. Behav.* 75, 651-659.

Pincemail, J., Bonjean, K., Cayeux, K., Defraigne, J.O. (2002). Physiological action of antioxidant defences. *Nutr. Clin. Metabol.* 16, 233-239.

Poli, G. (1993). Liver damage due to free radicals. *Br. Med. Bull.* 49 (3), 604-620.

Pousada, R.C., Cyrne, M.L., Hayes, D. (1979). Characterization of preribosomal ribonucleoprotein particles from *Tetrahymena pyriformis. Eur. J. Biochem.* 102 (2), 389-397.

Powers, S.K., Jackson, M.J. (2008). Exercise-induced oxidative stress: Cellular mechanisms and impact on muscle force production. *Physiol. Rev.* 88, 1243-1276.

Powers, S.K., Lennon, S.L. (1999). Analysis of cellular responses to free radicals: focus on exercise and skeletal muscle. *Proc. Nutr. Soc.* 58, 1025-1033.

Preiss, J., Kosuge, T. (1970). Regulation of enzyme activity in photosynthetic systems. *Ann. Rev. Plant. Physiol.* 21, 433-466.

Proust, B. (2006). Petite Gëomëtrie des Parfums. Editions du Seuil. Paris. 126.

Pupillo, P. (1972). The specificity of glyceraldehyde-3-phosphate dehydrogenase in green plants, Euglena and Ochromonas. *Phytochem.* 11, 153-161.

Puytorac, P.De., Batisse, A., Bohatier, J., Corliss, J.O., Deroux, G., Didier, P., Dragesco, J., Fryd-Versavel, G., Grain, J., Groliere, C.A., Hovasse, R., Iftode, F., Laval, M., Roque, M., Savoie, A., Tuffrau, M. (1974). Proposal for a classification of the phylum Ciliophora doflein, 1901. *C. R. Acad. Sci.* Paris, 278, 2799-2802.

Ralser, M., Wamelink, M.M., Kowald, A., Gerisch, B., Heeren, G., Struys, E.A., Klipp, E., Jakobs, C., Breitenbach, M., Lehrach, H., Krobitsch, S. (2007). Dynamic rerouting of the carbohydrate flux is key to counteracting oxidative stress. *J. Biol.* 6 (4), 10.

Repine, J. E., Bast, A., Lankhorst, I. (1997). Oxidative stress in chronic obstructive pulmonary disease. Oxidative Stress Study Group. *Am. J. Respir. Crit. Care Med.* 156, 341357.

Richard, H. (1992). Epices et Aromates. Technologie et Documentation Lavoisier. Paris. 339.

Robert, G. (2000). The Senses of Perfume. Osman Eroylles Multimedia. Paris. 224.

Robertson EF, Dannelly HK, Malloy PJ and Reeves HC (1987) Rapid isoelectric focusing in a vertical polyacrylamide minigel system. *Anal. Biochem.* 167, 290-294.

Roche, E., Romero-Alvira, D. (1996). Alteraciones del DNA inducidas por el estres oxidativo. *Med. Clin.* 106, 144-153.

Rogstam, A., Larsson, J.T., Kjelgaard, P., Wachenfeldt, C.V. (2007). Mechanisms of adaptation to nitrosative stress in *Bacillus subtilis*. *J. Bacteriol.* 189 (8), 3063-3071.

Ruzicka, L. (1953). The isoprene rule and the biogenesis of terpenic compounds. *Experientia.* 9, 357-396.

Ryrfeldt, A., Bammenberg, G., Moldius, P. (1993). Free radical and lung disease. *Br Med Bull.* 49 (3), 588-603.

Sachdev, S., Davies, K.J.A. (2008). Production, detection, and adaptive responses to free radicals in exercise. *Free Rad. Biol.* Med. 44, 215-223.

Sahoo, R., Sengupta, R., Ghosh, S. (2003). Nitrosative stress on yeast: inhibition of glyoxalase-I and glyceraldehyde-3-phosphate dehydrogenase in the presence of GSNO. *Biochem. Biophys. Res. Commun.* 302 (4), 665-670.

Salmon, A.B., Richardson, A., Përez, V.I. (2010). Update on the oxidative stress theory of aging: does oxidative stress play a role in aging or healthy aging? *Free Radic. Biol. Med.* 48, 642-655.

Samokyszyn, V. M., Marnett, L. J. (1990). Inhibition of liver microsomal lipid peroxidation by 13-cis-retinoic acid. *Free Radic. Biol. Med.* 8 (5), 491-496.

Satir, B.H., Wissig, S.L. (1982). Alveolar sacs of *Tetrahymena*: Ultrastructural characteristics ans similarities to subsurface cisterns of muscle and nerve. *J. Cell. Sci.* 55, 13-33.

Scherz-shouval, R., Elazar, Z. (2011). Regulation of autophagy by ROS: physiology and pathology. *Trends Biochem. Sci.* 36, 30-38.

Schlapfer, B.S., Zuber, H. (1992). Cloning and sequencing of the genes encoding glyceraldehyde-3-phosphate dehydrogenase, phosphoglycerate kinase and triosephosphate isomerase (gap operon) from mesophilic *Bacillus megaterium*: comparison with corresponding sequences from thermophilic *Bacillus stearothermophilus*. *Gene.* 122, 53-62.

Schuppe-Koistinen, I., Moldc'us, P., Bergman, T., Cotgreave, I.A. (1994). S-thiolation of human endothelial cell glyceraldehyde-3-phosphate dehydrogenase after hydrogen peroxide treatment. *Eur. J. Biochem.* 221 (3), 1033-1037.

Sen, N., Hara, M., Ahmed, A.S., Cascio, M.B., Kamiya, A., Ehmsen, J.T., Aggrawal, N., Hester, L., Dore, S., Snyder, S.H., Sawa, A. (2009). GOSPEL: A novel neuroprotective protein that binds to GAPDH upon S-nitrosylation. *Neuron.* 63 (1), 81-91.

Senatore, F. (1996). Influence of harvesting time on yield and composition of the essential oil of a thyme (Thymus pulegioides L.) growing wild in Campania (Southern Italy). *J. Agric. Food Chem.* 44, 1327-1332.

Serrano, A., Mateos, M. I., Losada, M. (1993). ATP-driven transhydrogenation and ionization of water in a reconstituted glyceraldehyde-3-phosphate dehydrogenase (phosphorylating and non-phosphorylating) model system. *Biochem. Biophys. Res. Commun.* 197(3), 1348-1356.

Sevanian, A., Hochstein, P. (1985). Mechanisms and consequences of lipid peroxidation in biological systems. *Annu. Rev. Nutr.* 5, 365-390.

Silva, J., Abebe, W., Sousa, S.M., Duarte, V.G., Machado, M.I.L., Matos, F.J.A. (2003). Analgesic and anti-inflammatory effects of essential oils of Eucalyptus. *J. Ethnopharmacol.* 89, 277-283.

Sirover, M.A. (2005). New nuclear functions of the glycolytic protein, glyceraldehyde-3-phosphate dehydrogenase, in mammalian cells. *J. Cell. Biochem.* 95, 45-52.

Slater, T. F. (1984). Free-radical mechanisms in tissue injury. *Biochem. J.* 222, 1-15.

Slaughter, J.C., Davies, D.D. (1968). Inhibition of 3-phosphoglycerate dehydrogenase by 1-serine. *Biochem. J.* 109, 749-755.

Sorg, O. (2004). Oxidative stress: a theoretical model or a biological reality? *C. R. Biol.* 327, 649-662.

Soukri, A., Hafid, N., Valverde, F., Elkebbaj, M.S., Serrano, A. (1996). Evidence for a posttranslational covalent modification of liver glyceraldehyde-3-phosphate dehydrogenase in hibernating jerboa (*Jaculus orientalis*). *Biochim. Biophys. Acta.* 1292, 177-187.

Stadtman, E.R., Berlett, B.S. (1998). Reactive oxygen-mediated protein oxidation in aging and disease. *Drug. Metab. Rev.* 30, 225-243.

Stamler, J.S., Singel, D.J., Loscalzo, L. (1992). Biochemistry of nitric oxide and its redox-activated forms. *Science.* 258, 1898-1902.

Stocker, R., Keaney, F. Jr (2004). Role of oxidative modifications in atherosclerosis. *Physiol. Rev.* 84, 1381-1478.

Sun, J.H., Xin, C.L., Eu, J.P., Stamler, J.S., Meissner, G. (2001). Cysteine-3635 is responsible for skeletal muscle ryanodine receptor modulation by NO. *Proc. Natl. Acad. Sci.* USA. *98*, 11158-11162.

Tamoi, M., Ishikawa, T., Takeda, T., Shigeoka, S. (1996). Enzymic and molecular characterization of NADP-dependent glyceraldehyde-3-phosphate dehydrogenase from *Synechococcus* PCC 7942: resistance of the enzyme to hydrogen peroxide. *Biochem. J.* 316, 685-690.

Tisdale, E.J. (2001). Glyceraldehyde-3-phosphate dehydrogenase is required for vesicular transport in the early secretory pathway. *J. Biol. Chem.* 276 (4), 2480-2486.

Tozlua, E., Cakirb, A., Kordalic, S., Tozluc, G., Ozerd, H., Akcine, T.A. (2011). Chemical compositions and insecticidal effects of essential oils isolated from Achillea gypsicola, Satureja hortensis, Origanum acutidens and Hypericum scabrum against broadbean weevil (Bruchus dentipes). *Sci Hort.* 130, 9-17.

Vacchi, C., Piccari, G.G., Pupillo, P. (1973). Characterization of NADP-linked glyceraldehyde-3-phosphate dehydrogenase of *Euglena gracilis*. *Z. Pflanzenphysiol.* 69, 351358.

Valko, M., Leibfritz, D., Moncol, J., Cronin, M.T.D., Mazur, M., Telser, J. (2007). Free radicals and antioxidants in normal physiological functions and human disease. *Biocell.* 39,

44-84.

Valko, M., Rhodes, C.J., Moncol, J., Izakovic, M. (2006). Free radicals, metals and antioxidants in oxidative stress-induced cancer. *Chemico-Biol. Interact.* 160, 1-40.

Valverde, F., Losada, M., Serrano, A. (1997). Functional complementation of an Escherichia coli gap mutant supports an amphibolic role for $NAD(P)^+$ -dependent glyceraldehyde-3-phosphate dehydrogenase of Synechocystis sp. Stran PCC 6803. *J. Bacteriol.* 179, 4513-4522.

Vertuani, S., Angusti, A., Manfredini, S. (2004). The antioxidants and pro-antioxidants network: an overview. *Curr. Pharm. Des.* 10 (14), 1677-1694.

Wassman, S., Wassman, K., Nichenig, G. (2004). Modulation of oxidant and antioxidant enzyme expression and function in vascular cells. *Hypertension.* 44, 381-386.

Wenqtang, G., Shufen, L., Ruixiang, Y., Shaokun, T., Can, Q. (2007). Comparison of essential oils of clove buds extracted with supercritical carbon dioxide and other three traditional extraction methods. *Food chem.* 1001, 1558-1564.

Wolfe, J. (1973). Conjugation in *Tetrahymena*. The relation between the division cycle and cell pairing. *Dev. Biol.* 35, 221-231.

Yonushot, G.R., Orthwerth, B.J., Koeppe, O.J. (1970). Purification and properties of a nicotinamide adenine dinucleotide phosphate requiring glyceraldehydes-3-phosphate dehydrogenase from spinach leaves. *J. Biol. Chem.* 245, 4193-4198.

Yu, B.P. (1994). Cellular defenses against damage from reactive oxygen species. *Physiol Rev.* 74 (1), 139-162.

Zabka M., Pavela R., Slezakova L. (2009). Antifungal effect of *Pimenta dioica* essential oil against dangerous pathogenic and toxinogenic fungi. *Ind. Crop. Prod.* 30, 250-253.

Zelko, I.N., Marian, T.J., Folz, R.J. (2002). Superoxide dismutase multigene family: a comparison of the CuZn-SOD (SOD1), Mn-SOD (SOD2) and EC-SOD (SOD3) gene structures, evolution and expression. *Free Rad. Biol. Med.* 33, 337-349.

Zeyuan, D., Bingyin, T., Xiaolin, L., Jinming, H., Yifeng, C. (1998). Effect of green tea and black tea on the blood glucose, the blood triglycerides and antioxidation in aged rats. *J. Agric. Food Chem.* 46(10), 3875-3878.

Zhang, J., Snyder, S.H. (1992). Nitric oxide stimulates auto-ADP-ribosylation of glyceraldehyde-3-phosphate dehydrogenase. *Proc. Natl. Acad. Sci. USA.* 89 (20), 9382-9385.

Zhao, Y., Nakashima, S., Andoh, M., Nozawa, Y. (1997). Cloning and sequencing of a cDNA encoding glyceraldehydes-3-phosphate dehydrogenase from *Tetrahymena thermophila*: Growth-associated changes in its mRNA expression. *J. Euk. Microbiol.* 44 (5), 434-437.

Buy your books fast and straightforward online - at one of world's fastest growing online book stores! Environmentally sound due to Print-on-Demand technologies.

Buy your books online at
www.morebooks.shop

Kaufen Sie Ihre Bücher schnell und unkompliziert online – auf einer der am schnellsten wachsenden Buchhandelsplattformen weltweit! Dank Print-On-Demand umwelt- und ressourcenschonend produziert.

Bücher schneller online kaufen
www.morebooks.shop